CW00508841

The little book of Shocking Eco-facts

FIELL

Published by Fiell Publishing Limited
www.fiell.com

A catalogue record for this book is available from the British Library.

ISBN 978-1-906863-12-8

Note: The publisher has endeavoured to ensure that the information
contained in this book was correct at the time of going to press.

Project Concept: Charlotte & Peter Fiell
Editor: Lee Ripley
Art Direction, Design, Picture Sourcing:
Jonathan Abbott and Daniel Street at Barnbrook
Design: Izzy Way at Barnbrook
Creative Direction: Jonathan Barnbrook

This book has been printed in China using soy inks — an environmentally
friendly alternative to traditional petroleum-based ink.

The little book of Shocking Eco-facts

Compiled by Mark Crundwell & Cameron Dunn

Chapter One: The Land: P.010

Chapter Two: The Oceans: P.066

Chapter Three: The Atmosphere: P.124

Comment / Reference: P.176

Introduction

Rachel Carson's book *Silent Spring*, published in 1962, is credited with being the springboard for the modern environmental movement. Controversial and compelling, *Silent Spring* awoke a blissfully ignorant world to the grave threat to wildlife and humans of DDT (dichloro-diphenyltrichloroethane), a widely used chemical insecticide developed in the 1950s against insect-borne diseases such as malaria and typhus. While effective, it was also destructive to the environment. Carson assessed the health of the land, oceans and atmosphere and argued that unless drastic steps were taken to reduce human misuse of the natural environment and seek biological alternatives, all too soon there would be a 'silent spring'.

Carson's theme of an impending environmental catastrophe was further developed in Hardin's 1968 article, 'The Tragedy of the Commons'. Linking damage to the environment with the surging global population, Hardin provided the inspiration behind the first World Earth Day in 1970 and the influential Club of Rome report published in 1972 called 'Limits to Growth'. In these early years, the focus of environmental concern was often on the impending shortage of resources that would be caused by the burgeoning population. The world had at last begun to realise that the earth's capacity for provision was not limitless and that continued unchecked consumption would lead to environmental disaster.

The essentially pessimistic outlook of the 1970s was transformed in the 1980s by the emergence of a new paradigm, that of sustainable development. Attention shifted from concern about the environment and resources to faith in the supremacy of the market, and the belief that continued economic growth could be compatible with protecting the environment. Influential works such as Bruntdland's *Sustainable Development* seemed to suggest that contrary to the Club of Rome report, there were no limits to growth. By adopting so-called 'green strategies' or 'green growth' we could effectively still have our cake and eat it. This was enthusiastically adopted by 'green' economists such as David Pearce in his books *World without End* and *The Economic Value of Biodiversity* (co-authored with Dominic Moran). This focus on economics instead of conservation resulted in other, perhaps less environmentally sensitive, economists such as Bederman (1996) and Lomberg (2001) challenging the assertions of the environmental movement and declaring that they thought things were not worse. Ecological concerns were apparently either wrong or over-hyped and if only the environment were left to the market, both people and the environment would prosper.

Since then, a decade of debt-fuelled economic expansion and the rapid development of the so called BRICs (Brazil, Russia, India and China) have once again focused our attention on what Ward and Dubos had pithily titled their book: *Only One Earth* (1972). The development of what have been termed 'ecological footprints' has firmly established that at our current rate of resource use, the human race is now living beyond the ability of the planet to support us. Well publicised extinctions such as the Baiji (the Yangtze Dolphin) in China have again awoken the world to the looming eco-catastrophe that Carson predicted nearly 50 years ago.

Slowly the attitude of the world is changing. To some extent scientific fact has caught up with rhetoric and anthropogenic (human caused) climate change is now being accepted as fact, not simply theory. The main polluters are at last recognising their role in the changes being inflicted on our atmosphere and ultimately on all the species that inhabit the earth, including *homo sapiens*. Many people are beginning to come to the same conclusion as Heinberg did, that far from a 'World without End', we are facing a century of declines.

As a result of this change in attitude, over a billion dollars has been raised to save what are arguably the most environmentally important areas of the planet — biodiversity hotspots. Indeed, some of the most endangered species on earth, such as the kakapo parrot, are now being protected in predator-free sanctuaries and are starting the slow road to recovery. For other species such as the Pinta Island tortoise of the Galapagos, it is too late. Lonesome George is the last of his kind — a testament to how slowly we are learning the lessons of trying to balance development and conservation.

This book highlights the current state of the planet 50 years on from when Rachel Carson first warned the public about the potential for environmental disaster if steps were not taken to stop the damage we were causing. We hope that these shocking eco facts will serve as ammunition that will help people push decision makers into adopting policies and strategies that will allow both people and the environment to find a course that will ensure that there will be no future 'silent spring'.

This book is therefore divided into three parts — examining the impact of people on **THE LAND**, **THE OCEANS** and **THE ATMOSPHERE**.

THE LAND

Humans must have somewhere to live and today over 6.6 billion of us crowd the land and ferociously consume its resources. Terrestrial biodiversity has not coped well with the onslaught of human expansion. No surprise then that as pressure to consume resources has grown, biodiversity, hotspots have been identified — the most ecologically important areas of the planet. These biodiversity hotspots, which are unique small islands and tropical forests, already face grave threats. Their continued survival is vital, therefore, in order to preserve a pool of crucial genetic resources. The impact of agriculture and the destruction of habitats has affected many species both directly and indirectly. We are living beyond the ability of the planet to support our current resource consumption. The facts are there and this section looks at the impact on the earth beneath our feet.

THE OCEANS

Covering some 70% of the planet's surface, the oceans and seas seem almost limitless in their extent and potential to provide resources for humans. Two problems have emerged in the last few decades however. First, the realisation that the oceans are interconnected which means pollution and waste spread inexorably to pollute the whole rather than one part. Similarly ocean-going trade has helped to connect the world as never before, but it also spreads alien invasive species to new areas, often with disastrous consequences. Secondly, there is now recognition that the limitless nature of the oceans' bounty was illusory. We are increasingly aware that the oceans have been over-exploited and degraded.

THE ATMOSPHERE

Our atmosphere envelops the earth acting as a protective, life-sustaining blanket. This tenuous layer provides the air we breathe, regulates the temperature at the earth's surface and protects life from harmful ultraviolet radiation. Yet human activity continues to pollute the atmosphere, changing its chemistry and potentially altering the earth's climate. Despite many words written on the dangers of global warming, we who inhabit this planet have yet to act in a meaningful way to stop, or even slow down, the process.

Having examined global problems, more regional and local issues must be considered. While we may have been able to make some progress on acid deposition by banning the burning of coal in some areas or looking at cleaner ways of burning fuel — the reality is somewhat less encouraging. The impact of local air pollution in urban areas, another issue that was widely thought to have been solved by the Clean Air Acts of the 1950s, is still a major threat to the health of those living in urban areas. Even in our own homes, the damage we are creating to our environment pursues us and we face the growing problem of indoor air pollution.

The purpose of this book is to act as a wakeup call, rather than to present a litany of depressing facts and statistics. Every sorry eco-fact has a solution which we humans are capable of putting in place. Many of the solutions are simple. Almost all require a change of attitude towards the air we breathe, the water we drink or the land we live on. The phrase 'think global, act local' may have become a cliché, but it is as true today as it always has been. Consider your own impact on the planet and change your attitudes and actions to protect it from further degradation. Write a protest letter. Buy local produce. Cycle to work. Recycle at home. Numerous small, local changes can and will generate a more fundamental global change.

Epilogue

At the time of going to press, the scale of the environmental
disaster caused by the oil spill in the Gulf of Mexico
was still being assessed. The impact on the marshlands,
the wildlife and the livelihoods of the people in
Louisiana and the surrounding states will certainly be
felt for many years and some ecosystems may never recover.
This underscores more than ever the need for action
to be taken to save the global environment.

The Land

95%

*The total percentage of life
that existed on this planet
and is now extinct.*

10,000

*The current estimated rate of
extinctions per year, compared to
the long term rate of 1,000 per year.*

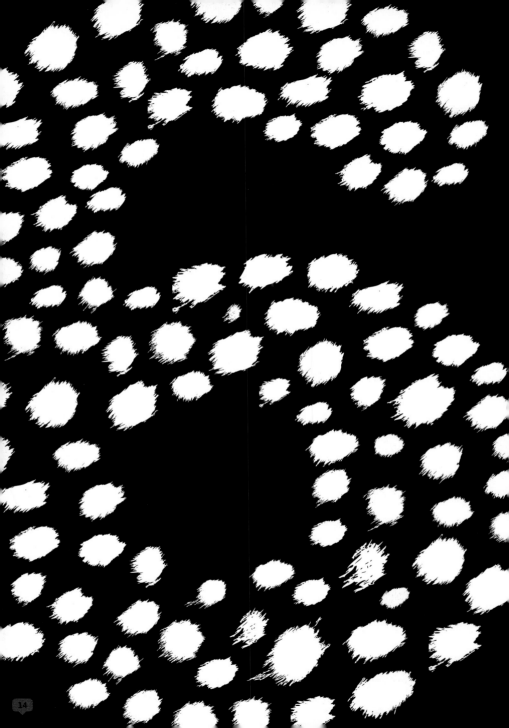

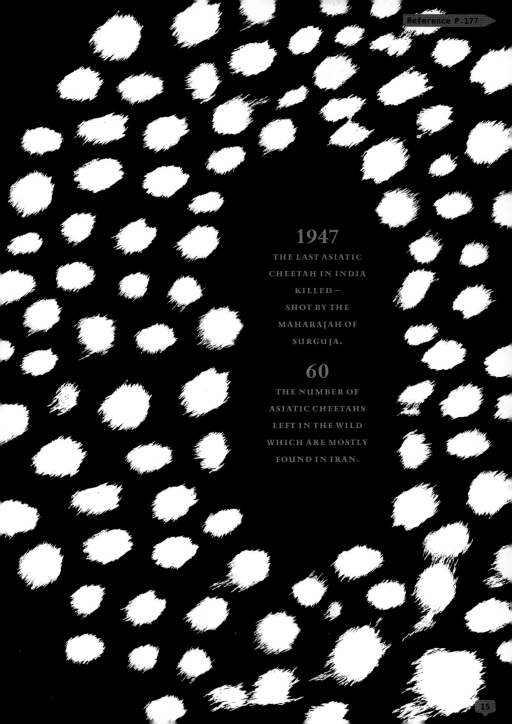

1947
THE LAST ASIATIC
CHEETAH IN INDIA
KILLED—
SHOT BY THE
MAHARAJAH OF
SURGUJA.

60
THE NUMBER OF
ASIATIC CHEETAHS
LEFT IN THE WILD
WHICH ARE MOSTLY
FOUND IN IRAN.

$327

The cost of an illegally traded orang-utan sold as a pet in the wildlife markets in Indonesia.

8

The number of orang-utans killed on average to secure the one sold in to the illegal wildlife market.

0

The number of Baiji

(the Yangtze Dolphin) *left —*

the aquatic mammal was declared extinct in 2007.

2,000

The remaining population

of the close relative to the Baiji —

the Ganges Dolphin

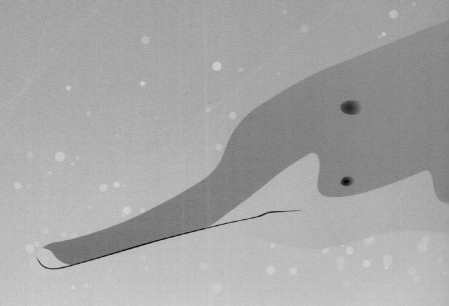

2.3%

The total area of the earth covered by 34 conservation hotspots which contain nearly 50% of the world's plant species and 42% of the world's bird, mammal, reptile and amphibian species.

The amount of the native vegetation already lost in these conservation hotspots.

70%

52%

The percentage of globally threatened mammals that are hotspot endemics.

The predicted loss of such endemic species given an extra 1,000 square kilometres of habitat loss.

47%

99.8%

The percentage of primary forest cover
that Singapore has lost since colonisation

62%

The percentage of epiphytic species such as orchids
that have been lost in the same period along with
26% of all the native species of Singapore and the
entire rich mangrove epiphyte flora

Reference P.178

25

The total number of 'alalas left in either
captivity or the wild. The introduction of
avian pox and avian malaria by sailors
into the south Kona region of Hawaii
has resulted in 'alalas becoming the
most endangered member of the
crow family (corvid) in the world.

The number of Pinta giant tortoises left in the world. Despite all attempts, Lonesome George appears to be doomed to extinction.

The number of goats removed from Isabela Island in the Galapagos in the hope of preventing the fate of Lonesome George happening to the island's unique giant tortoises.

100,000

9,704

The total number of endemic plant species in Madagascar.

4,323

The predicted number of future extinctions of native
or endemic plants in Madagascar meaning that 45%
of all endemic plant species will be lost with the
removal of another 1,000 km² of rainforest.

124

The remaining number of kakapo parrots which are endemic to New Zealand.

0

The number of moas left after only 100 years of human settlement of New Zealand.

6%

The percentage of the world's surface area covered in tropical rainforest.

75% The total percentage of the earth's species of plants and animals that are found in tropical rainforest.

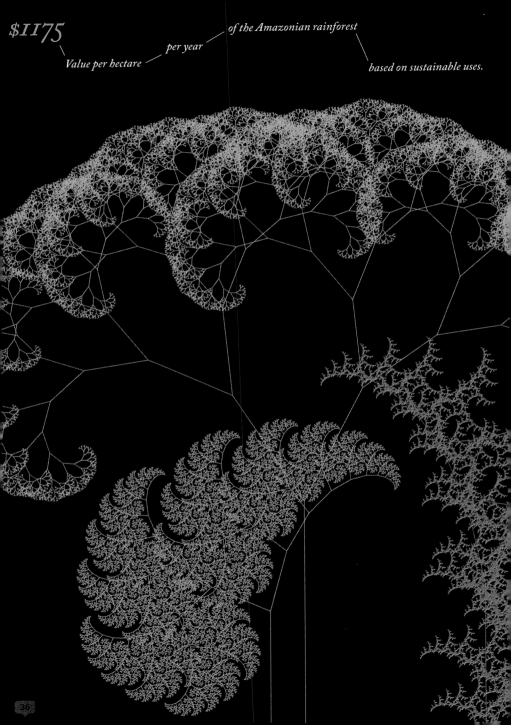

$*II75*$

Value per hectare

per year

of the Amazonian rainforest

based on sustainable uses.

$110—150
Value per hectare — per year
if converted to pasture
or other land use.

7,140

FOOTBALL PITCH. THE AREA IN SQUARE METRES OF AN INTERNATIONAL

THE NUMBER OF SECONDS YOU HAVE TO COUNT TO BEFORE ANOTHER FOOTBALL PITCH SIZED AREA OF TROPICAL RAINFOREST IS DESTROYED.

1.5

40 MILLION TONNES of greenhouse gases were emitted in 2007 from the mining of the Alberta Tar Sands. This is equivalent to the pollution of ten million cars.

142 MILLION TONNES: The estimated amount of greenhouse gases which would be emitted by 2020 if the growth of the Tar Sands extraction is left unchecked.

32,000 SQUARE MILES: The land area already approved by the Alberta government for exploration and development by the big oil companies.

60 MILLION: THE ESTIMATED NUMBER OF BISON LIVING IN THE GREAT PLAINS OF AMERICA BEFORE THE EUROPEAN COLONISATION OF THE 19TH CENTURY.

1,000: THE ESTIMATED NUMBER REMAINING AFTER THE GREAT SLAUGHTERS OF 1870–1873 AND 1880–1883.

1/3 } ———————————————— THE NUMBER OF BIRD SPECIES THAT HAVE BECOME EXTINCT IN THE MOUNTAIN FORESTS OF SAN ANTONIO IN THE COLOMBIAN ANDES.

80 } —————— THE NUMBER OF YEARS IN WHICH THIS EXTINCTION
HAS TAKEN PLACE.

1/3 The amount of forest cover lost on Mt. Kilimanjaro in the last 70 years.

25% The percentage of this forest cover that is irreplaceable upper montane and subalpine forest.
The destruction has been caused by fires induced by changes in the climate.

*The number of prescribed drugs that have their
base in products found in tropical forests.*

The total value of medical products based on products found in tropical forests.

4 tonnes of top soil for each woman, man and child on the planet that is lost every year due to soil erosion.

Reference P.180

10 million hectares of arable land per year that is being lost due to soil erosion.

80: The amount of nitrogen
(in Teragrams Nitrogen/Year) emitted into
the atmosphere in the form of ammonia
(NH₃) caused by human activity.
[1 teragram = 1,000,000,000 kilograms].

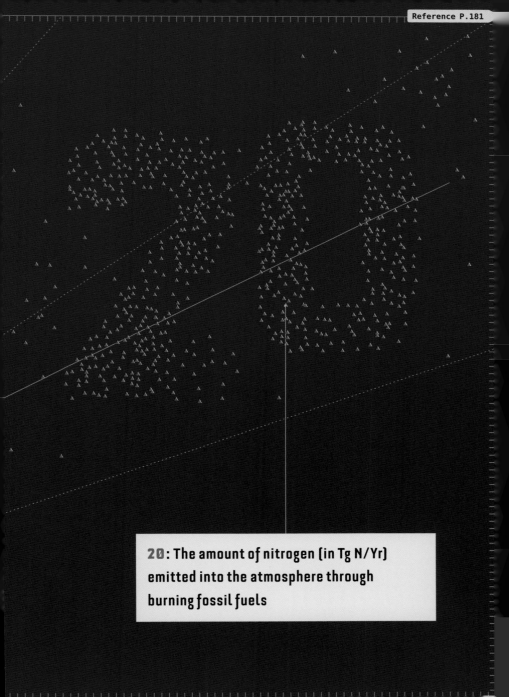

20: The amount of nitrogen (in Tg N/Yr) emitted into the atmosphere through burning fossil fuels

The amount of applied nitrogen fertilizer that is taken up by crops.

The reduction in the nitrogen efficiency of crops between 1960 and 2000.

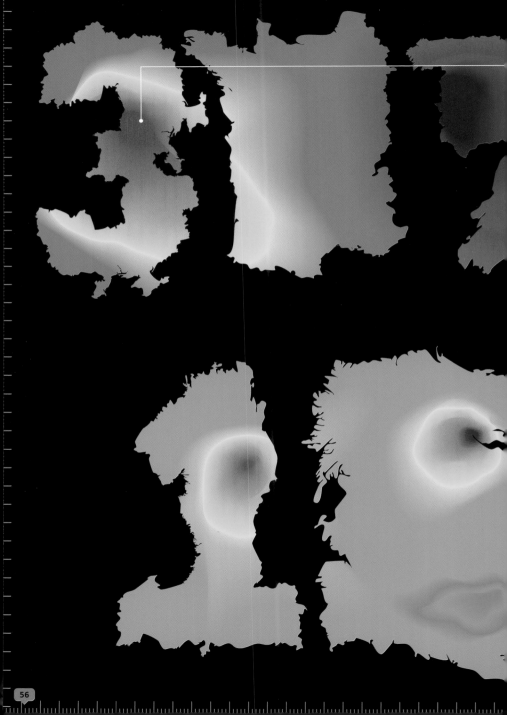

The decline in common farmland birds in Europe over the period 1978 to 2002.

The increase in the use of pesticides over the same period.

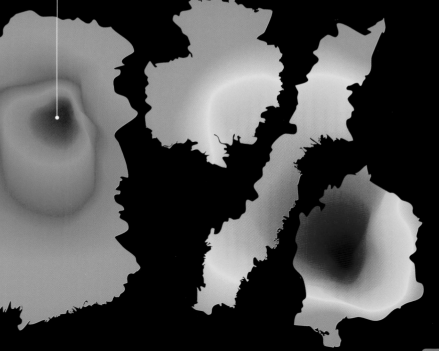

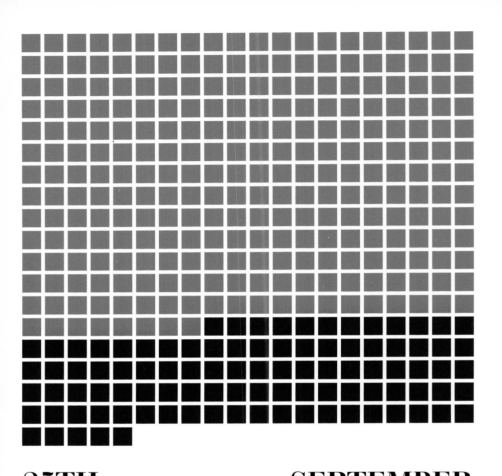

25TH SEPTEMBER

THE DAY WHEN THE WORLD
GOES INTO ECOLOGICAL
DEBT IN TERMS OF ITS
RESOURCE CONSUMPTION
AND WASTE DISPOSAL.

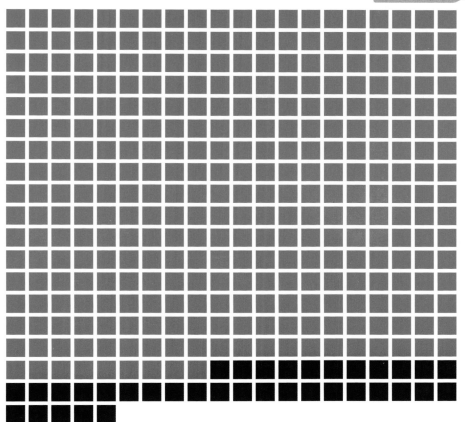

25TH NOVEMBER

THE EQUIVALENT DEBT DAY BACK IN 1995 HIGHLIGHTING THE FACT THAT WE ARE OVER-USING OUR RESOURCES AT AN EVER INCREASING PACE.

3 out of the **9** critical natural
boundaries for earth
that have already been passed.

These three are

climate change,
rate of biodiversity loss
and interference in the nitrogen cycle.

4

The number of the remaining

6 critical natural boundaries

that are in imminent danger of being passed:

global freshwater use,

change in land use,

ocean acidification,

and interference with the global phosphorus cycle.

Reference P.181

The number of global hectares per capita that the United States of America needs to maintain its present consumption pattern.

94

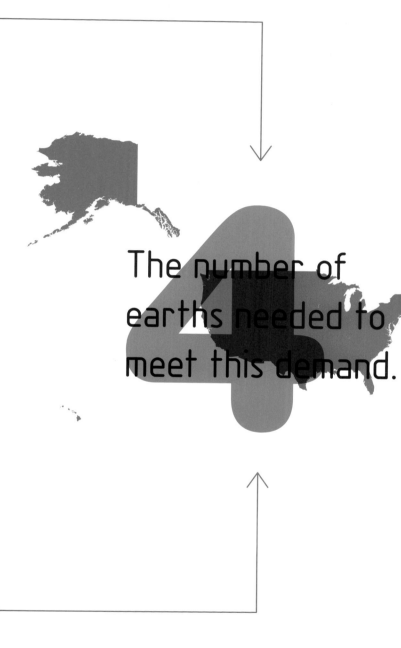

The number of
earths needed to
meet this demand.

—One—
The number of species (humans) using
86.6% of the world's biocapacity.

—*1.7 million*—
The number of known species using the
remaining 13.4% of the world's biocapacity.

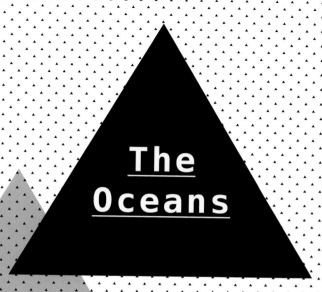

The Oceans

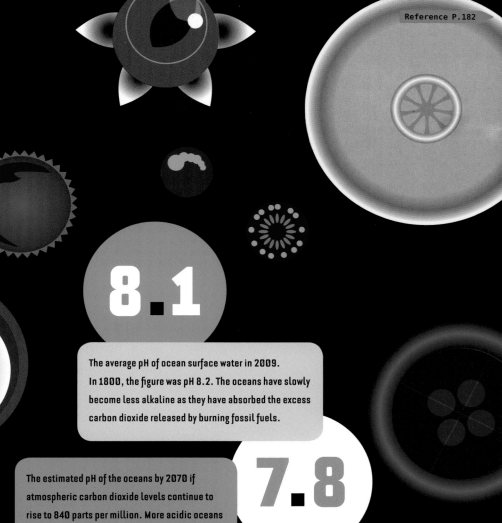

8.1

The average pH of ocean surface water in 2009. In 1800, the figure was pH 8.2. The oceans have slowly become less alkaline as they have absorbed the excess carbon dioxide released by burning fossil fuels.

7.8

The estimated pH of the oceans by 2070 if atmospheric carbon dioxide levels continue to rise to 840 parts per million. More acidic oceans will threaten the life-cycles of coral and many other marine organisms.

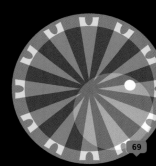

1 MILLION TONNES PER HOUR

THE AMOUNT OF ANTHROPOGENIC CARBON DIOXIDE BEING ABSORBED BY THE OCEANS.

5,000–10,000 YEARS

THE TIMESCALE FOR NATURAL CHANGES TO OCEAN ACIDITY OF A SIMILAR MAGNITUDE TO THAT CAUSED BY HUMANS IN ONLY 50-100 YEARS.

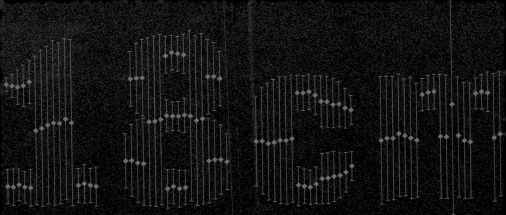

The global rise in sea level between 1900 and 2000, based on a rise of 1.8mm per year.

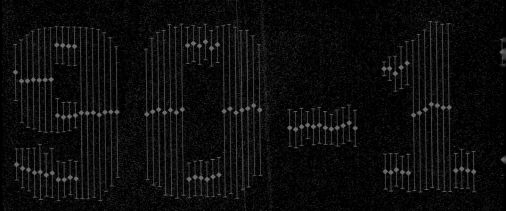

The estimated sea level rise between 2000 and 2100. Sea levels will rise due to thermal expansion as the oceans continue to warm, plus additional water from melting ice caps, glaciers and permafrost.

Reference P.182

1.5 – 2.0°C: The rise in sea temperatures that results in coral reef bleaching events, which leads to degradation and death of reefs.

2.8°C: The average projected increase in global temperatures by 2099.

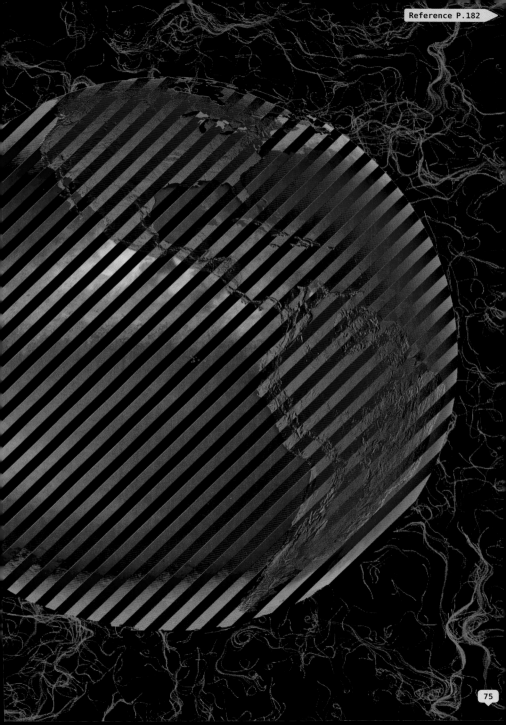

85% OF THE WORLD'S MAJOR RIVER DELTAS

ARE SINKING

ACCORDING TO A STUDY PUBLISHED IN 2009

50%: PROJECTED INCREASE

IN THE AREA OF DELTA LAND THAT

WILL BE FLOODED BY 2050

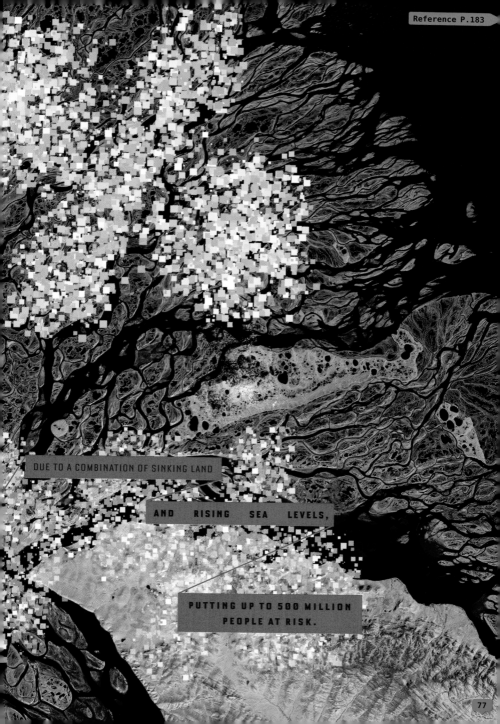

DUE TO A COMBINATION OF SINKING LAND

AND RISING SEA LEVELS,

PUTTING UP TO 500 MILLION
PEOPLE AT RISK.

-1°C to -3°C

The cooling the Northwest of Europe could experience if the warm ocean currents from the Gulf Stream and North Atlantic Drift collapsed due to global warming. The Norwegian coast might experience a cooling of up to -12°C.

Reference P.183

51 Small Island Developing States ^(SIDS) are especially vulnerable to **climate change** and **sea level rise.**

Many are **low lying,** isolated and depend heavily on the oceans for resources.

Reference P.183

0.02%
The proportion of **global greenhouse gas emissions**
from the 51 SIDS.

865,000

One estimate of the combined population of Minke, Humpback and Fin whales in the North Atlantic prior to the onset of commercial whaling.

215,000

Estimate of the combined population of these whales since the 1990s.

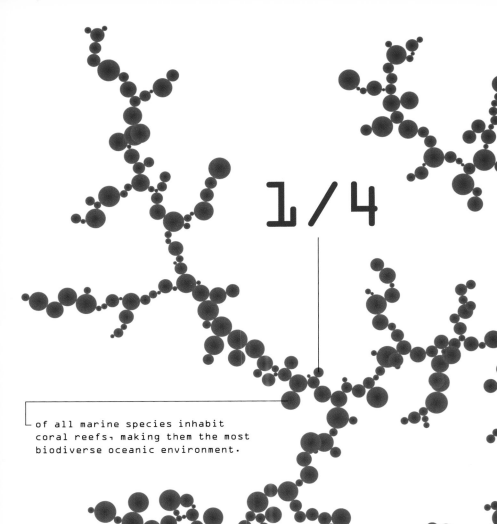

1/4

of all marine species inhabit coral reefs, making them the most biodiverse oceanic environment.

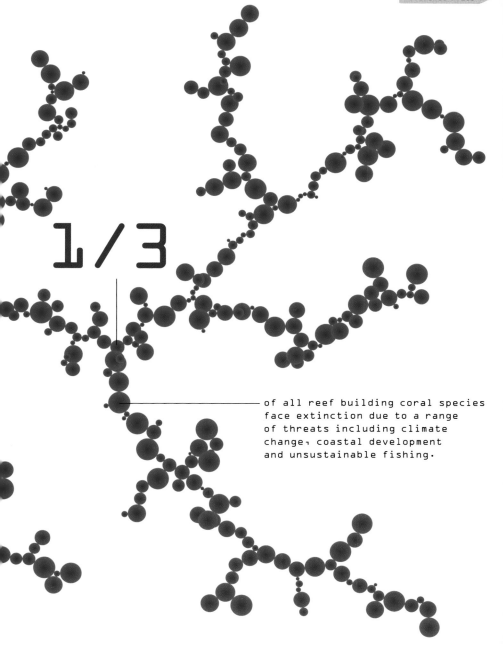

1/3

of all reef building coral species face extinction due to a range of threats including climate change, coastal development and unsustainable fishing.

35 MILLION TONNES ___ The global wild fish catch in 1960 when world population stood at 3 billion.

92 MILLION TONNES ___ The global wild fish catch in 2006 with a world population continuing to rise past 6.5 billion.

11/18

11 of the 18 species of penguin show evidence of declining populations.

7

A further 7 penguin species are classified as vulnerable.

4

4 penguin species are endangered.

250 PARTS PER M
OF PCB CONTA
SOME NORTHERN
WHALES. 1 PART
THE LEVEL OF PCB
FOUND IN MOST

Reference P.184

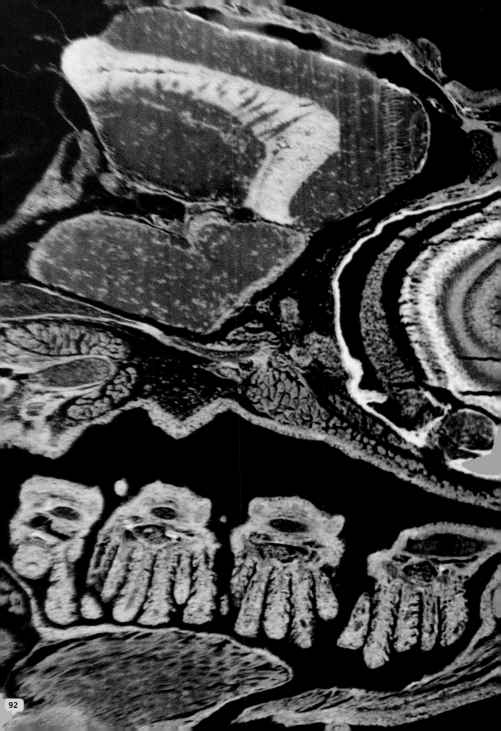

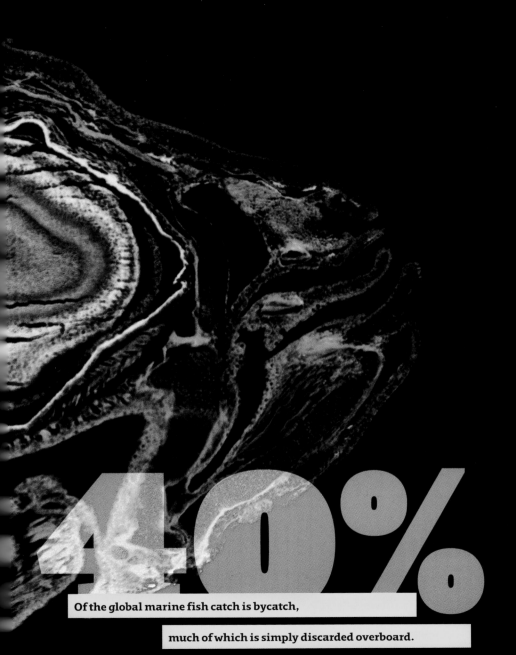

40%

Of the global marine fish catch is bycatch,

much of which is simply discarded overboard.

2048

One

marine

global

have

year

collapsed

meaning

by

the

oceans

as

a

useful

9

number

billion

of

The

in

at

rates

of

2048,

current

Reference P.184

estimate

of

the

stocks

will

fish

90%,

effectively

end

of

food.

source

of

wild

to

the

mouths

feed

global

growth

population

50% The percentage of the world's wild fish which are caught in an area covering only 7.5% of the world's oceans, meaning that food resources from the oceans are highly geographically concentrated.

9.9 kilograms
The worldwide average annual consumption of fish per person in the 1960s

16.7 kilograms
The worldwide average annual consumption of fish in 2006 per person. About 1 billion people rely almost solely on fish as a source of protein.

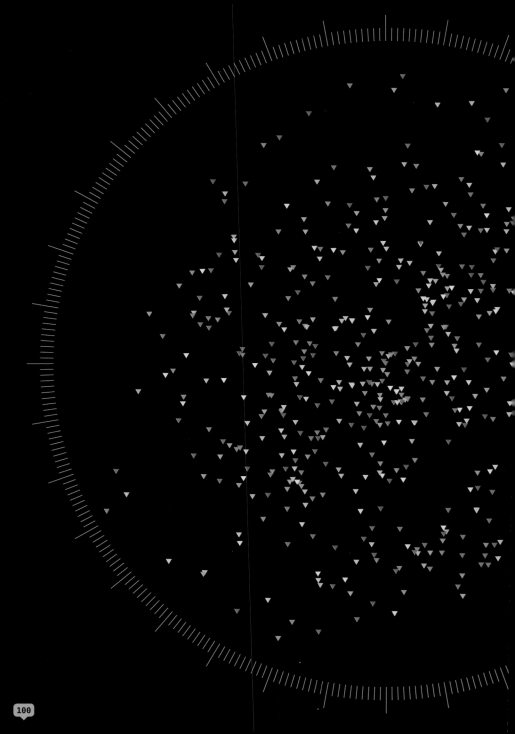

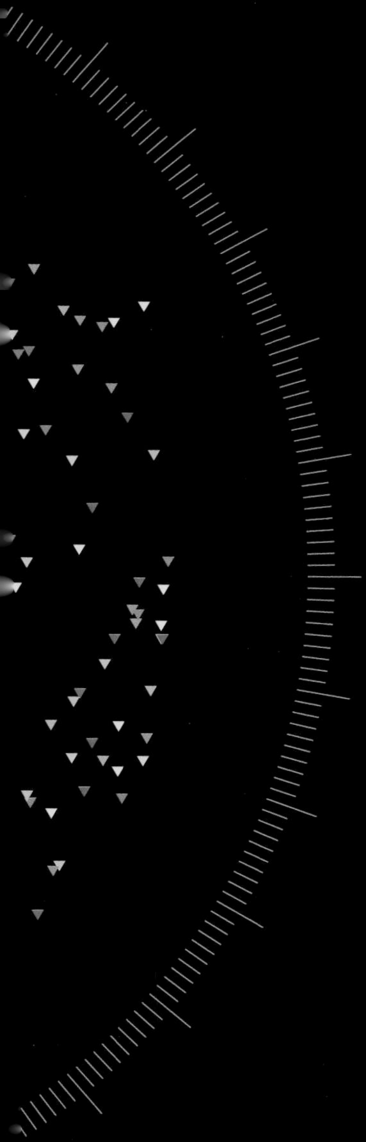

1% of the world's
fleet of 3.5 million
fishing boats are
large, factory
fishing vessels.

50% of the world's
annual marine fish
catch of 80-90
million tonnes is
caught by large,
factory fishing
vessels.

200,000 loggerhead sea turtles and 50,000 leatherbacks caught up in fishing gear worldwide.

Populations of both species have fallen by 80-90% in the past decade.

20%
The percentage of coastal mangrove forests which have been destroyed worldwide since 1980, amounting to a loss of 3.6 million hectares of mangrove.

Mangroves are critically endangered or approaching extinction in 26 of the 120 countries where mangrove forest is found.

Reference P.185

20
MILLION TONNES

The amount of wild marine fish used every year to produce fish food for farmed fish.

5
KILOGRAMS

The number of kilograms of wild marine fish needed to produce food to feed 1 kilogram of farmed salmon to maturity.

100 MILLION TONNES

Of floating rubbish is believed to be trapped in the Great Pacific Garbage Patch, much of it plastic waste. Vast circular ocean currents, or gyres, sweep rubbish into continental scale waste patches.

18,000
Pieces of plastic trash in
every square kilometre of
ocean.

1.3 million tonnes

The amount of oil discharged into the sea, worldwide, each year: Equivalent to 3 of the world's largest super-tankers discharging their entire cargo. (This figure doesn't take into account the 4.9 million barrels released into the sea by the Gulf of Mexico oil spill of 2010).

14%

Decline in the global marine Living Planet Index 1970-2005.

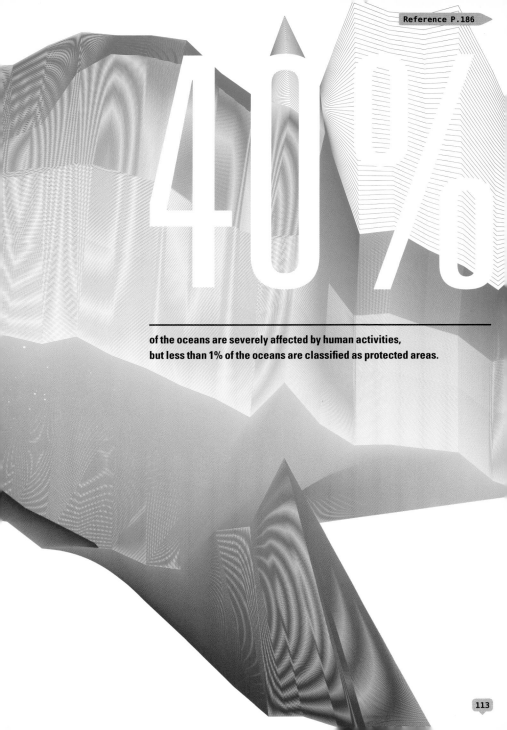

40%

of the oceans are severely affected by human activities, but less than 1% of the oceans are classified as protected areas.

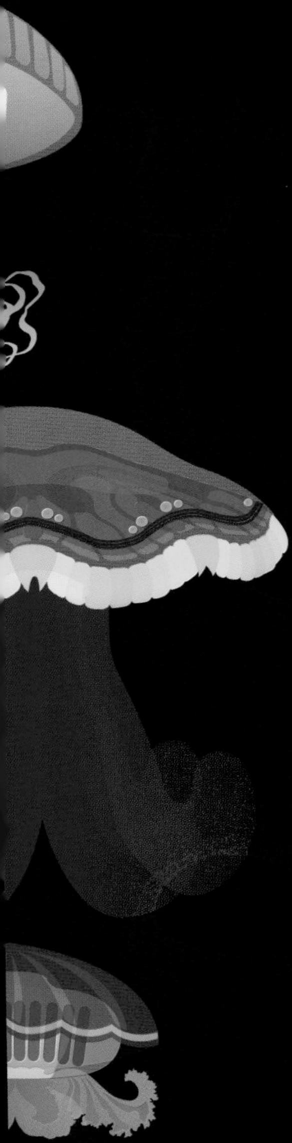

33%

Plastic bags
and other plastic debris
account for one third of all
deaths of leatherback
sea turtles.

405

The number of dead zones

reported ⋯⋯ in the world's oceans

in 2008.

These are areas

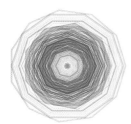

so depleted of oxygen

they can no longer support life.

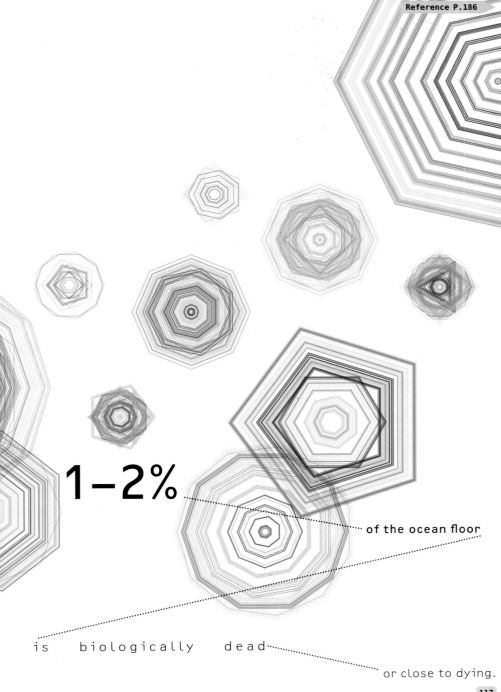

1-2%

of the ocean floor

is   biologically   dead

or close to dying.

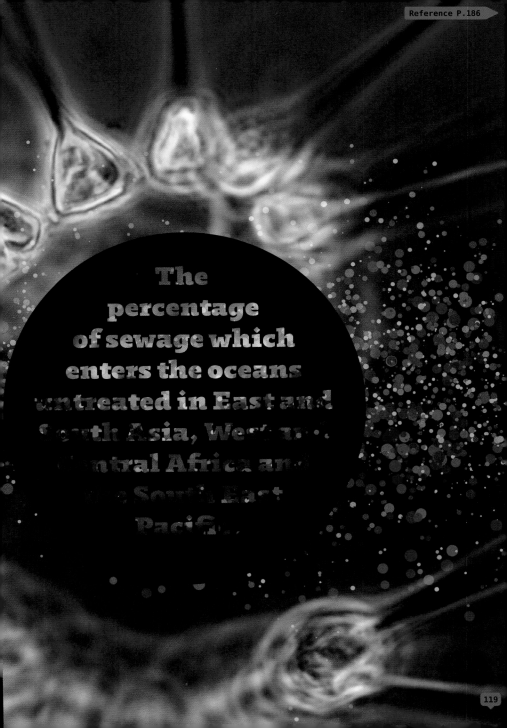

The percentage of sewage which enters the oceans untreated in East and South Asia, West and Central Africa and the South East Pacific...

3 0 %
THE INCREASE IN MERCURY LEVELS IN THE PACIFIC OCEAN BETWEEN THE MID 1990S AND 2006.

<u>5</u>   <u>0</u>   <u>%</u>
THE   FURTHER
INCREASE   IN
MERCURY   LEVELS
IN   THE   PACIFIC
EXPECTED BY 2050.

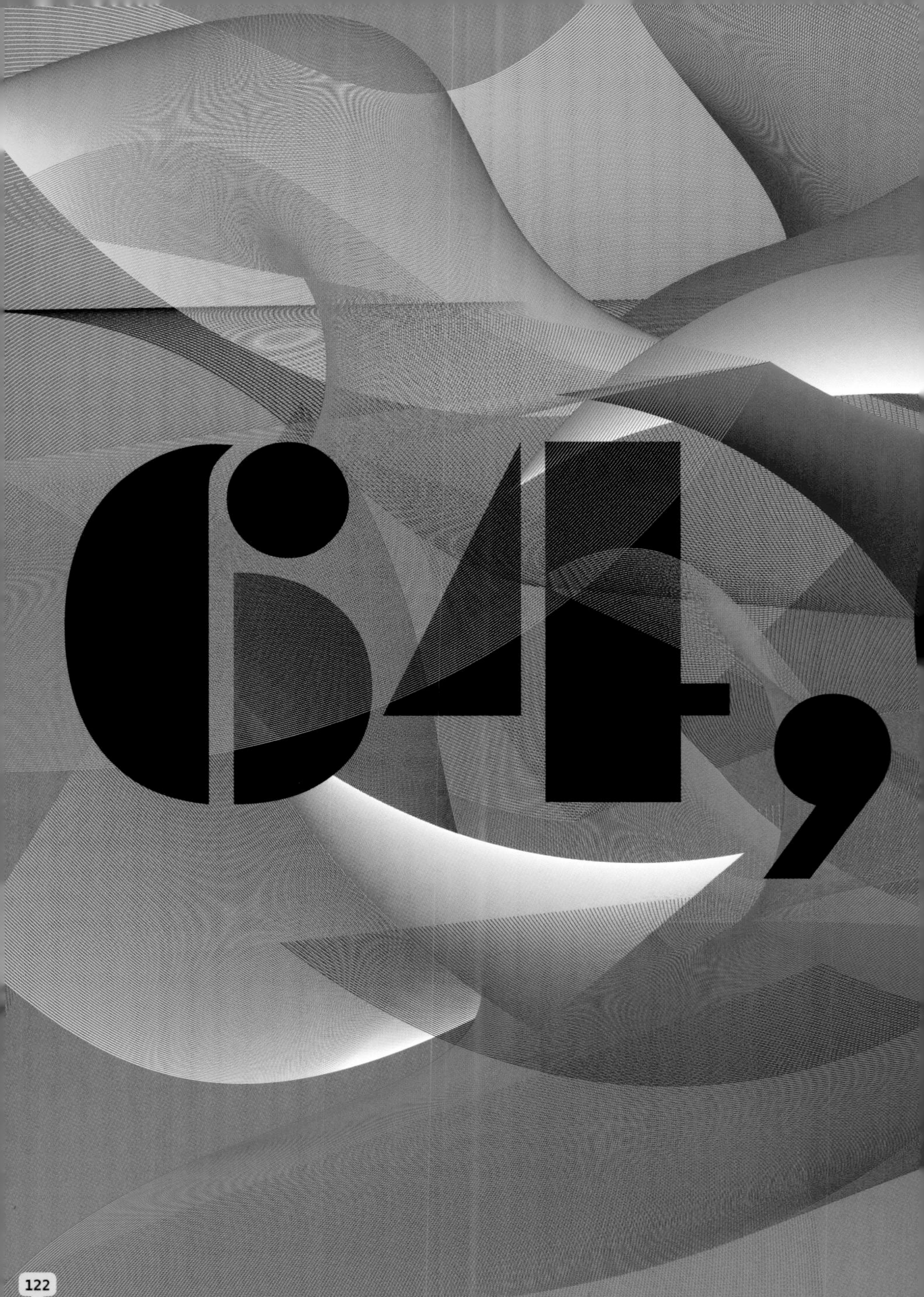
64,

The number of ships of over 400 tonnes which traversed the world's oceans in 2007, emitting 1019 million tonnes of carbon dioxide

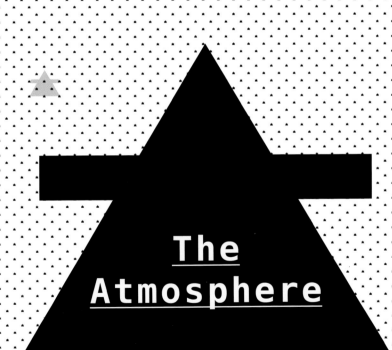

The Atmosphere

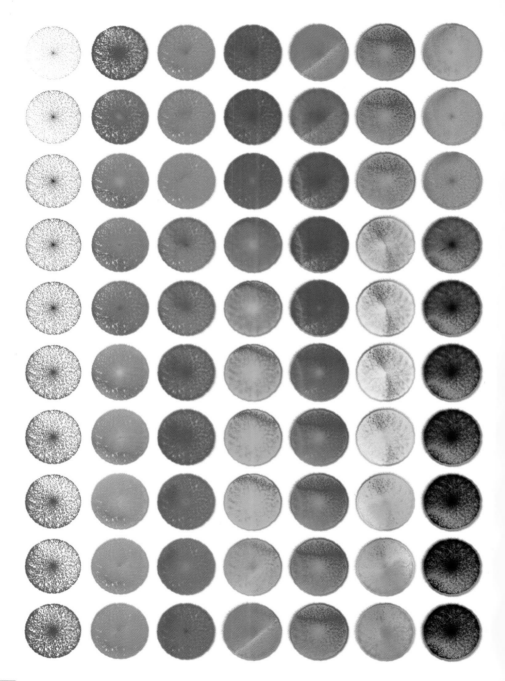

**87 PARTS PER MILLION.
THE CONCENTRATION
IN 2009, AN INCREASE
OF 23% IN 50 YEARS.
CARBON DIOXIDE LEVELS
ARE CONSIDERED TO BE
HIGHER TODAY THAN AT
ANY TIME IN THE LAST
650,000 YEARS.**

**315 PARTS PER MILLION.
THE CONCENTRATION OF
CARBON DIOXIDE IN THE
ATMOSPHERE IN 1958.
THIS WAS THE
YEAR CONTINUOUS
MEASUREMENT OF CARBON
DIOXIDE LEVELS BEGAN AT
MAUNA LOA ON HAWAII.**

450

parts per million of carbon dioxide has been identified
by some scientists as a 'tipping point' for climate change,
beyond which change would be dangerous and irreversible.

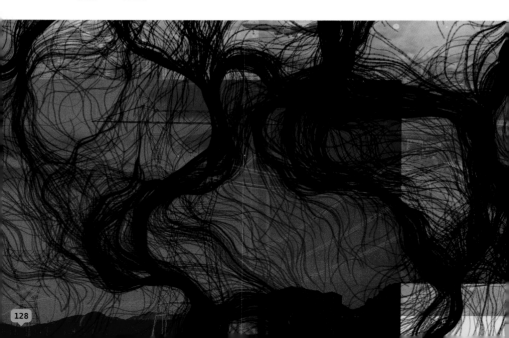

2042

The approximate year, at current rates, that the world will reach a concentration of 450 parts per million of carbon dioxide.

0.02 *Tonnes of carbon dioxide emitted per person by the average Somali and Afghani in 2006. Most African and many Asian nations emitted under 1 tonne of carbon dioxide per person.*

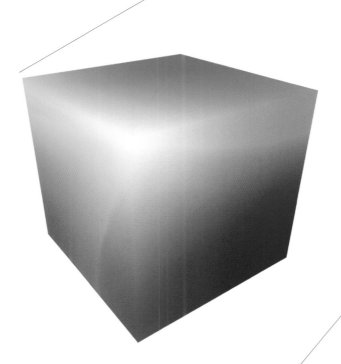

56.2 *Tonnes of carbon dioxide per person emitted by the average Qatari, the highest level in the world in 2006. Developed and oil-rich nations emit 100–1000 times more carbon dioxide per person than poorer developing nations.*

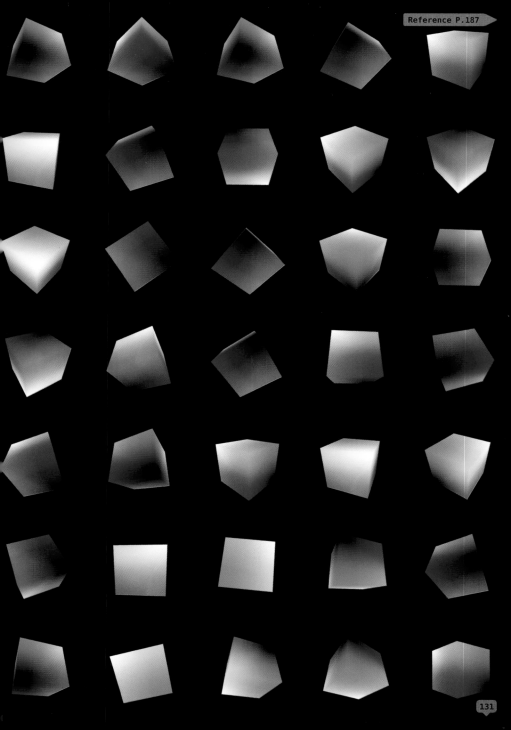

148%

The increase in concentration of methane levels in the atmosphere since 1750. The pre-industrial level of 715 parts per billion had risen to 1774 ppb by 2005.

Methane has 21 times the global warming potential of carbon dioxide, making it a much more powerful greenhouse gas.

75%
The average reduction
in electricity consumption
for lighting achieved by
switching incandescent bulbs
to more efficient CFL
or LED lighting.

40 Billion tonnes of carbon dioxide are expected to be released due to deforestation between 2008 and 2012.

*120—400
Tonnes is the
amount of carbon
stored by a single
hectare of
tropical forest.*

158 Million people in
live in areas where l
failed one or more na
standards. In 2007 th
US population.

he United States
al air quality
onal air quality
: was 52% of the

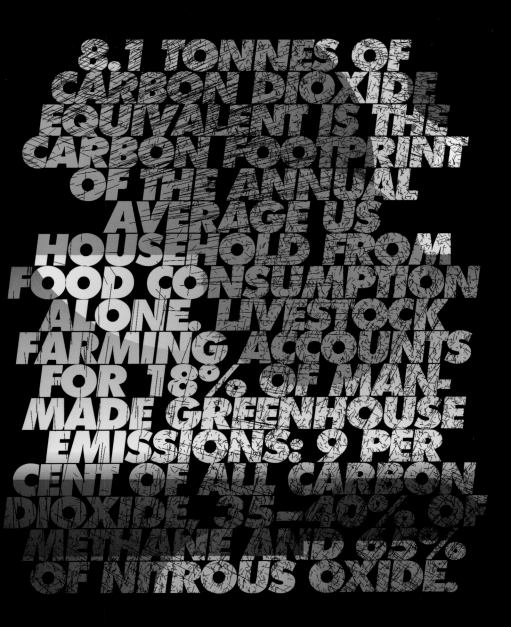

8.1 TONNES OF CARBON DIOXIDE EQUIVALENT IS THE CARBON FOOTPRINT OF THE ANNUAL AVERAGE US HOUSEHOLD FROM FOOD CONSUMPTION ALONE. LIVESTOCK FARMING ACCOUNTS FOR 18% OF MAN-MADE GREENHOUSE EMISSIONS: 9 PER CENT OF ALL CARBON DIOXIDE, 35–40% OF METHANE AND 65% OF NITROUS OXIDE.

2050 THE EARLIEST THAT THE GLOBAL OZONE LAYER COULD RECOVER FROM DEPLETION CAUSED BY CFCS AND OTHER POLLUTANTS.

1987 THE YEAR THE MONTREAL PROTOCOL ON SUBSTANCES THAT DEPLETE THE OZONE LAYER WAS SIGNED.

20% Decline in sunlight since the 1970s at Guangzhou in China, as a result of rising levels of industrial pollution.

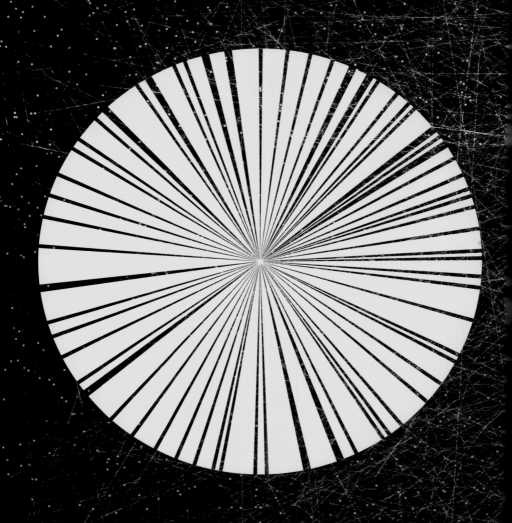

3–4% Decline in sunlight over much of India and China, per decade, since the 1950s.

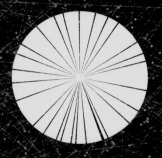

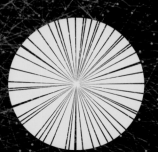
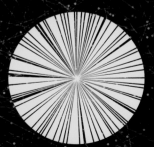

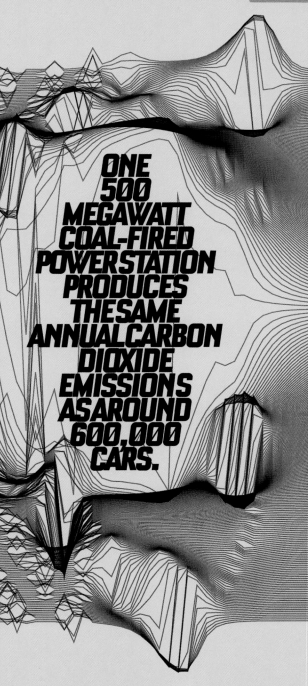

ONE 500 MEGAWATT COAL-FIRED POWER STATION PRODUCES THE SAME ANNUAL CARBON DIOXIDE EMISSIONS AS AROUND 600,000 CARS.

Sulphur Dioxide, Carbon Dioxide, Nitrogen Oxide, Carbon Monoxide, particulate matter, unburned hydrocarbons, ash, arsenic, lead, cadmium and mercury are all released into the atmosphere when coal is burnt.

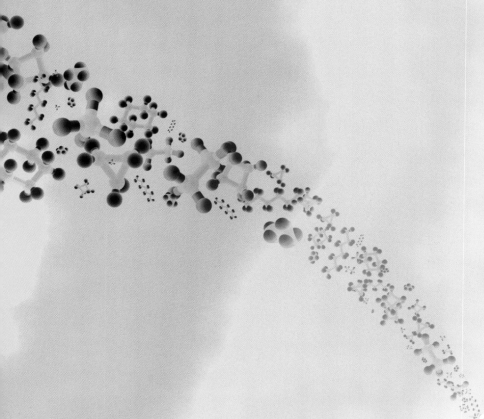

The global warming potential of HFC-134a,
a gas used as an alternative to ozone depleting CFCs.

38

340%

30

The increase in concentration of HFC-134a
in the atmosphere between 1998 and 2005.

THE →

APPROXIMATE →

THE ←

PERCENTAGE →

4 AND 5 ←

TROPICAL →

CYCLONES →

75

INCREASE →

← IN

OF →

← CATEGORY

SINCE →

1970.

153

3–4 °C

The warming experienced in Western Canadian and Alaskan Arctic over the last 50 years. The Arctic has warmed twice as fast as the global average in the past few decades and this trend is expected to continue in the 21st century.

Reference P.189

23% THE DROP IN NUMBER OF SPECIES IN GRASSLANDS IN THE UK CAUSED BY NITROGEN DEPOSITION.

17 THE AVERAGE DEPOSITION RATE
OF NITROGEN IN KG N HA-1 YR -1
(KILOGRAMS OF NITROGEN PER
HECTARE PER YEAR) OVER THE UK
FROM BOTH AGRICULTURAL AND
ENERGY PRODUCTION SOURCES.

40% THE PERCENT- AGE OF OAK STANDS SHOWING SEVERE DEFOLIATION IN GERMANY.

25% THE
PERCENT-
AGE OF ALL
TREES IN 27
EUROPEAN
COUNTRIES
THAT ALSO
SHOWED
SEVERE
DEFOLIATION

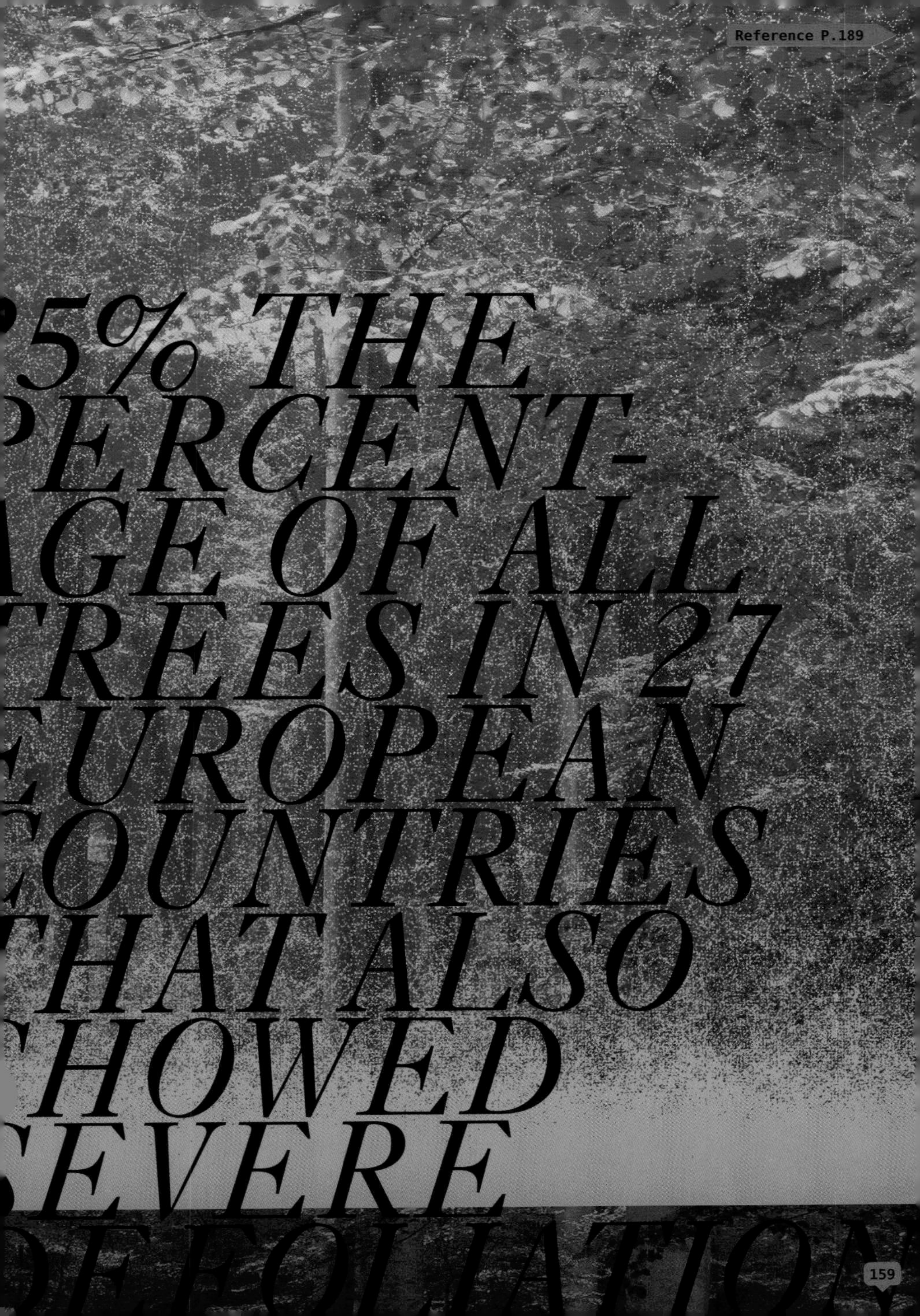

316,000
The number
of lost years
due
to people dying
prematurely
due to air pollution
in cities
in France in
one year.

55%
The number
of these
premature
deaths
attributable
to air
pollution
caused by
transport.

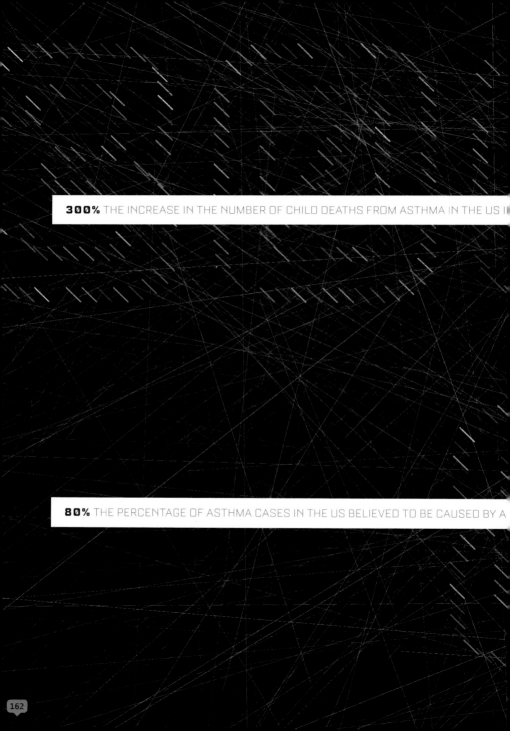

300% THE INCREASE IN THE NUMBER OF CHILD DEATHS FROM ASTHMA IN THE US I

80% THE PERCENTAGE OF ASTHMA CASES IN THE US BELIEVED TO BE CAUSED BY A

Reference P.190

OD 1979 TO 1996.

TO ALLERGENS AND INDOOR POLLUTION.

4,000
THE
NUMBER
OF
TOXIC
CHEMICALS
IN
ENVIRONMENTAL
TOBACCO
SMOKE.

30% INCREASE IN THE RISK OF DEATH FROM HEART DISEASE DUE TO PASSIVE SMOKING.

The number of cigarettes that you would need to smoke a day to equal the average air pollution in Mexico City.

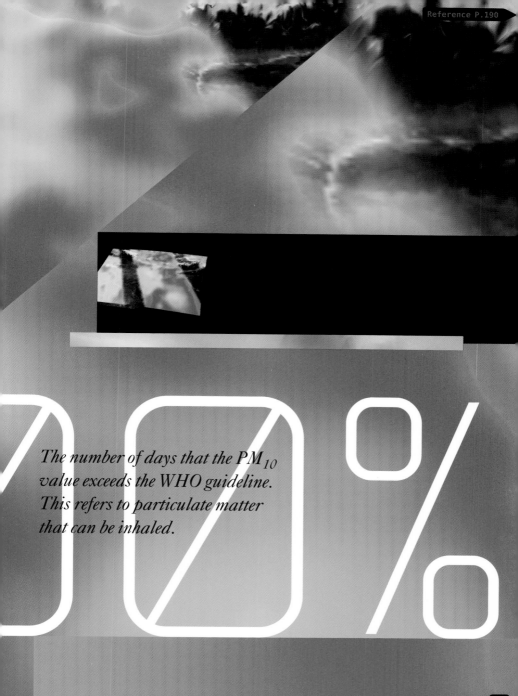

The number of days that the PM$_{10}$ value exceeds the WHO guideline. This refers to particulate matter that can be inhaled.

1.4 MILLION

The number of premature d e a t h s every year in Africa caused by the use of d i r t y biomass fuels to meet families' daily e n e r g y needs.

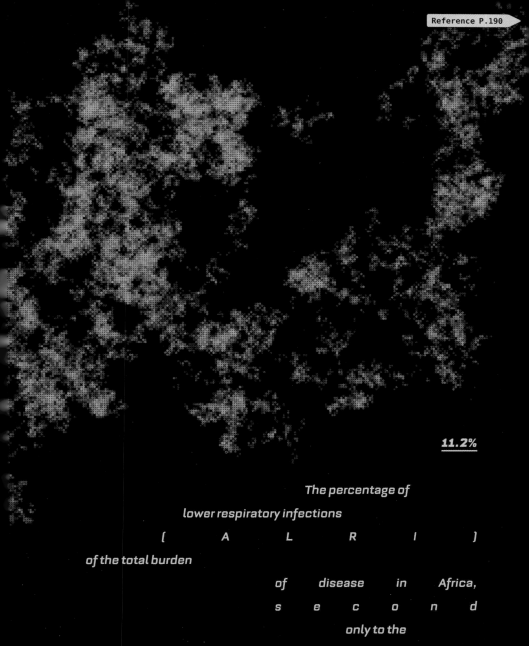

11.2%

The percentage of
lower respiratory infections
[A L R I]
of the total burden
of disease in Africa,
s e c o n d
only to the
continent's HIV/AIDS pandemic.

$8

The average cost of a clean cook stove sold in
Ethiopia that reduces the CO produced by such indoor stoves
by 76% and so below the WHO guidelines for good health.

$114

The average GDP per capita of Ethiopia in 2009.
This effectively means that it is unaffordable by
the vast majority of families in Ethiopia.

18,44

The number of → *registered residential vehicles →*

RNIA,

GE

AN

TION

PACITY

EN'S

OBTAIN

$4.5

ATER

DAMAGE

SIVE

8,200

in California

IN CALIFOR
THE DAMAG
FROM URBA
AIR POLLUT
TO THE CAP/
OF CHILDRI
LUNGS TO (
OXYGEN IS
TIMES GREA
THAN THE D
FROM PASS
SMOKING.

(105 PPB)
The maximum
suggested short-term
concentration of NO2 that
can be tolerated before
there are severe health
consequences.

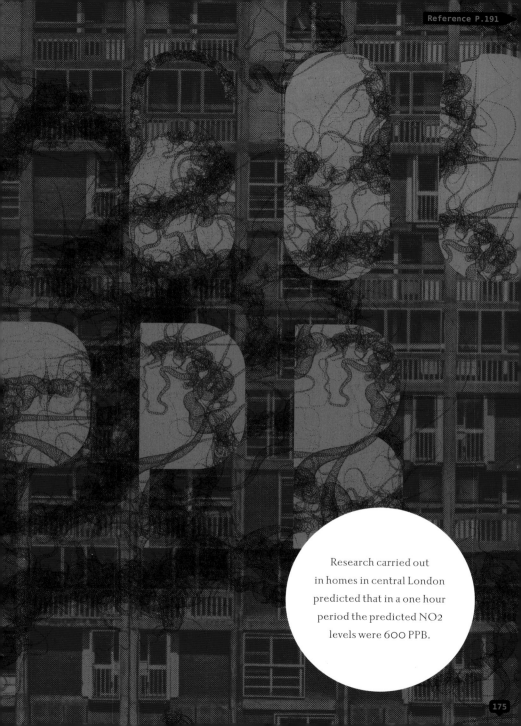

Research carried out
in homes in central London
predicted that in a one hour
period the predicted NO2
levels were 600 PPB.

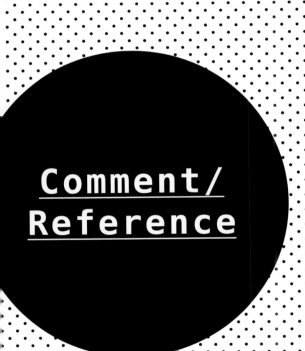

Comment/
Reference

Introduction

REFERENCE: Beckerman, W. (1996) *Small is Stupid: Blowing the Whistle on the Greens*, London: Gerald Duckworth. || Carson, R. (1962) *Silent Spring*, New York: Houghton Mifflin. || Hardin, G. (1968) 'The Tragedy of the Commons' *Science*, 162.3859, pp, 1243 – 1248. http://www.sciencemag.org/cgi/content/full/162/3859/1243 || Heinberg, R. (2007) *Peak Everything: Waking Up to the Century of Declines*, London: New Society Publishers; FEP Torn edition. || Lomborg, B. (2001) *The Skeptical Environmentalist: Measuring the Real State of the World*, Cambridge: Cambridge University Press. || Lomborg, B. (2007) *Cool it: The Skeptical Environmentalist's Guide to Global Warming*, New York: Knopf. || Meadows, D. and Meadows, D. (1972) *Limits to Growth: A Report for the Club of Rome's Project on the Predicament of Mankind*, New York: Earth Island. || Pearce, D. and Moran, D. (2004) *The Economic Value of Biodiversity*, London: Earthscan. || Pearce, D. and Warford, J. (1993) *World without End: Economics, Environment and Sustainable Development*, Oxford: Oxford University Press. || Ward, B. and Dubos, R. (1972) *Only One Earth: The Care and Maintenance of a Small Planet*, New York: W. W. Norton and Company. || World Commission on Environment and Development (1987) *Our Common Future*, New York : Oxford University Press.

Chapter One: The Land
P. 12–13 | LAND FACT 1

95%: The total percentage of life that existed on this planet and is now extinct.

10,000: The current estimated rate of extinctions per year, compared to the long term rate of 1,000 per year.

COMMENT: The process of natural selection will inevitably cause some species to become extinct while other new species will evolve through the process of evolution. Species are becoming extinct at a rate that has not been seen since the last global mass-extinction event around 65 million years ago. The fossil record shows that the background extinction rate for marine life is 0.1–1 extinctions per million species per year; for mammals it is 0.2–0.5 extinctions per million species per year. It has been estimated that the rate of extinction of species to be 100 to 1,000 times greater than what would be considered natural. These extinctions are principally caused by changes in land use such as the development of agriculture and urbanization as well as the introduction of invasive species.

REFERENCE: Mace, G., Masundire, H. and Baillie, J. (2005) ' Biodiversity' in *Ecosystems and Human Wellbeing: Current State and Trends* H. Hassan, R. Scholes and N. Ash (eds.) http://www.maWeb.org/documents/document.273.aspx.pdf || Rockström, J., Steffen,W. , Noone, K., *et al.* (2009) 'Planetary boundaries: exploring the safe operating space for humanity' *Ecology and Society*, 14(2):32. || http://www.ecologyandsociety.org/vol14/iss2/art32/ || Sala, O., Chapin III, F., Armesto, J., *et al.* (2000) 'Global biodiversity scenarios for the year 2100' *Science*, 287, pp. 1770–1774.

P. 14–15 | LAND FACT 2

1947: The last Asiatic cheetah in India killed—shot by the Maharajah of Surguja.

60: The number of Asiatic cheetahs left in the wild which are mostly found in Iran.

COMMENT: Although hunting, shooting and fishing have resulted in the extinction of several species notably the dodo and Pinta Island giant tortoises, the main threat to top carnivores such as cheetahs and tigers is not hunting by people but the loss of habitat. These top carnivores typically require large areas in which to hunt. This loss of habitat is mainly due to land use changes such as agriculture and urbanisation.

REFERENCE: Mace, G., Masundire, H., and Baillie, J. (2005) 'Biodiversity' in *Ecosystems and Human Wellbeing: Current State and Trends* H. Hassan, R. Scholes and N. Ash (eds.) http://www.maWeb.org/documents/document.273.aspx.pdf

P. 16–17 | LAND FACT 3

$327: The cost of an illegally traded orang-utan sold as a pet in the wildlife markets in Indonesia.

8: The number of orang-utans killed on average to secure the one sold in to the illegal wildlife market.

COMMENT: The trade in wildlife can be seen as at the heart of the relationship between biodiversity conservation and sustainable development. Poor people derive financial benefits by selling what they see as 'their resource' while richer nations, already responsible for the mass extinctions of their own biodiversity, criticize these poorer countries for trading in threatened species. Primates such as the orang-utan are some of the most traded wildlife species in the world. Of the 17 species of apes (gibbons, orang-utans, chimpanzees and gorillas) recognised by the IUCN (2009), three are listed as vulnerable, seven as endangered and two as critically endangered including the orang-utan. Habitat loss is the number one threat, but many of the additional threats are a direct result of habitat loss. To secure one orang-utan the mother has to be killed first and often the infant clinging to her will die in the process or die later from mistreatment or stress. Orang-utans used to be found in ten regions, but are now found on only two islands, Sumatra in Indonesia and Borneo in Malaysia.

REFERENCE: Cowlishaw G. and Dunbar, R. (2000) *Primate Conservation Biology*, London and Chicago: University of Chicago Press. || IUCN (2009) 'IUCN Red List of Threatened Species', http:// www.iucnredlist.org || Nijman,V. (2009) 'An assessment of trade in gibbons and orang-utans in Sumatra, Indonesia', TRAFFIC Southeast Asia, Petaling Jaya, Selangor, Malaysia. || http://www.assets.panda.org/downloads/orang_report_06_04_09.pdf

P. 18–19 | LAND FACT 4

0: The number of Baiji (the Yangtze Dolphin) left – the aquatic mammal was declared extinct in 2007.

2,000: The remaining population of the close relative to the Baiji – the Ganges Dolphin.

COMMENT: Little is known about the majority of fish populations, but the global decline of commercially important fish stocks or populations is relatively well documented . Nearly 850 species have become extinct or at least extinct in the wild since 1500. The IUCN Red List provides the most extensive data on trends in species richness using the application of categories and criteria that are based on assessments of extinction risk. However, the IUCN Red List needs to be interpreted with caution, because for most taxonomic groups the assessments are incomplete and heavily biased toward the inclusion of the most threatened species.

REFERENCE: Baillie, J., Hilton-Taylor, C. and Stuart, S. (2009) *2008 IUCN Red List of Threatened Species: A Global Species Assessment.* IUCN, Gland, Switzerland. http://www.iucnredlist.org || Mace, G. and Lande, R. (1991) 'Assessing extinction threats: toward a reevaluation of IUCN threatened species categories' *Conservation Biology*, 5, pp. 148–157. || Mace, G., Masundire, H. and Baillie, J. (2005)' Biodiversity' in *Ecosystems and Human Wellbeing: Current State and Trends* H. Hassan, R. Scholes and N. Ash (eds.) http://www.maWeb.org/documents/document.273.aspx.pdf || Hutchings, J. and Reynolds, J. (2004) 'Marine fish stock population collapses: consequences for recovery and extinction risk' *BioScience*, 54.4, pp. 297–309. http://www.caliber.ucpress.net/dio/abs/10.164/0006-3568(2004)054[0297:MFPCCF]2.0.CO%3B2 || Ransom Myers' Stock Recruitment Database has data on trends for over 600 fish populations covering more than 100 species www.mscs.dal.ca/~myers/welcome.html

P. 20–21 | LAND FACT 5

2.3%: The total area of the earth covered by 34 conservation hotspots which contain nearly 50% of the world's plant species and 42% of the world's bird, mammal, reptile and amphibian species.

70%: The amount of the native vegetation already lost in these conservation hotspots.

COMMENT: The concept of biodiversity hotspots was originated by Myers in 1988. To qualify as a biodiversity hotspot a region must meet two strict criteria: it must contain at least 0.5% or 1,500 species of vascular plants which are endemic (found only in this place) and to have lost at least 70% of its original habitat.

REFERENCE: Myers, N., Mittermeier, C., Mittermeier, G., *et al.* (2000) 'Biodiversity hotspots for conservation priorities' *Nature*, 403, 853–858 || Mittermeier, R.A., Giol, P. R., Hoffman, M., *et al.* (2005) 'Hotspots revisited: earth's biologically richest and most endangered ecoregions' Conservation International, Arlington, VA. www.biodiversityhotspots.org

P. 22–23 | LAND FACT 6

52%: The percentage of globally threatened mammals that are hotspot endemics.

47%: The predicted loss of such endemic species given an extra 1,000 square kilometres of habitat loss.

COMMENT: Nature seems to have put many of her eggs into one basket, because these hotspots also contain what are termed endemic species—species which are only found in one locality in the world. Yet these hotspots have not only lost 70% of their native vegetation, but also suffer from some of the highest losses of habitat due to high human population growth rates and population densities. Although rates of habitat loss within these hotspots vary, these rates are accelerating and making the future of many of these endemic species looks bleak.

REFERENCE: Brooks, T., Mittermeier, R., Mittermeier, C., *et al.* (2002) 'Habitat loss and extinction in the hotspots of biodiversity' *Conservation Biology*, 16.4, pp. 909–923.

P. 24–25 | LAND FACT 7

99.8%: The percentage of primary forest cover that Singapore has lost since colonisation.

62%: The percentage of epiphytic species such as orchids that have been lost in the same period along with 26% of all the native species of Singapore and the entire rich mangrove epiphyte flora.

COMMENT: As with many small islands with unique habitats, Singapore is liable to suffer from species loss. Deforestation and disturbance have been the main cause of plant species extinction in Singapore. Yet despite this, many more species continue to survive than would be predicted given the 99.8% loss of primary forest. Even with such high rates of deforestation, small fragmented areas can play a vital role in maintaining biodiversity and should, therefore, be the focus of conservation efforts instead of adopting management strategies that resign themselves to the loss of species.

REFERENCE: Turner, I., Tan, H., Wee, Y., *et al.* (1994) 'A study of plant species extinction in Singapore: lessons for the conservation of tropical biodiversity' *Conservation Biology*, 8, pp. 705–712.

P. 26–27 | LAND FACT 8

25: The total number of 'alalas left in either captivity or the wild. The introduction of avian pox and avian malaria by sailors into the south Kona region of Hawaii has resulted in 'alalas becoming the most endangered member of the crow family (corvid) in the world.

COMMENT: In common with other islands, Hawaii has been vulnerable to the effects of both invasive species and in this case invasive diseases introduced by humans. Living on an isolated island, endemic species such as the 'alala have not developed immunity to diseases that are prevalent in mainland communities. Similar invasive diseases have also had deleterious impacts on the numbers of giant tortoises in the Galapagos Islands.

REFERENCE: Work, T., Massey, J., Rideout, B., *et al.* (1999) ' Fatal toxoplasmosis in free-ranging endangered 'alala from Hawaii' *Journal of Wildlife Diseases*, 36.2, 2000, pp. 205–212. http://www.jwildlifedis.org/cgi/reprint/36/2/2005

P. 28–29 | LAND FACT 9

1: The number of Pinta giant tortoises left in the world. Despite all attempts, Lonesome George appears to be doomed to extinction

100,000: The number of goats removed from Isabella Island in the Galapagos in the hope of preventing the fate of Lonesome George happening to the island's unique giant tortoises.

COMMENT: In many ways the Galapagos Islands represent the problems of maintaining small island endemism. Compared to Ecuador, to which the islands belong, the residents of the Galapagos have a higher standard of living with 0.56 health professionals per 100 people in the Galapagos compared to 0.19 in Ecuador, almost double the amount of schooling and a higher employment rate with 45% of the Galapagos population economically active compared to only 32.0 % of those in Ecuador.

As a result there has been a rapid population growth of the populated islands from under 1,000 in 1950 to over 12,000 in 2009. This inevitably increases the number of invasive species such as goats and puts pressure on other natural resources leading to the irreplaceable loss of endemic species.

REFERENCE: Honey, M. (2008) *Ecotourism and Sustainable Development: Who Owns Paradise?* 2e, Washington, D.C.: Island Press. || Stewart P. (2006) *Galapagos: The Islands That Changed the World*, London: BBC Books. || D'Orso, M. (2003) *Plundering Paradise: The Hand of Man on the Galapagos Islands*, London : HarperCollins.

P. 30–31 | LAND FACT 10

9,704: The total number of endemic plant species in Madagascar.

4,323: The predicted number of future extinctions of native or endemic plants in Madagascar meaning that 45% of all endemic plant species will be lost with the removal of another 1,000 km² of rainforest.

COMMENT: Endemic plants can be vital. The only known cure for childhood leukaemia, the Madagascar periwinkle, is one such endemic plant threatened with extinction in Madagascar: 70% of the habitat of the periwinkle has already been lost.

REFERENCE: Brooks, T., Mittermeier, R., Mittermeier, C., *et al.* (2002) 'Habitat loss and extinction in the hotspots of biodiversity' *Conservation Biology*, 16.4, pp. 909–923. || Food and Agriculture Organization (FAO) (2009) 'State of the world's forests 2009' FAO, Rome, Italy. http://www.fao.org/docrep/011/i0350e/i0350e.00.HTML || Oldfield, S., Lusty, C. and MacKinven, A. (1998) *The World List of Threatened Trees*, Cambridge, U.K.: World Conservation Press. || Walter, K. and Gillett, H. (1998) *IUCN Red List of threatened plants*, World Conservation Union, Gland, Switzerland.

P. 32–33 | LAND FACT 11

124: The remaining number of kakapo parrots which are endemic to New Zealand.

0: The number of moas left after only 100 years of human settlement of New Zealand.

COMMENT: New Zealand along with other islands has developed a remarkable range of endemic species. Most famous of these are the flightless birds such as the kiwi, the moa and the kakapo. These birds evolved to live in the forests of New Zealand safe from any top mammal carnivores which were absent from the island. The moa grew to be over six feet tall. Over a period of 100 years between the 13th and 14th centuries, immediately after the first human settlement in New Zealand, the moa was hunted to extinction. Unlike the moa, the threats to the kakapo came from invasive species. The colonisation of the island brought stoats, cats, rats and dogs, and as a result the species plummeted towards extinction. In 1995, there were only 50 known kakapo surviving on a handful of small island sanctuaries. As a result of an intensive conservation project, however, there are now 124 kakapo surviving on three islands. It is hoped that soon they may be re-introduced to the main islands of New Zealand.

REFERENCE: Jay, M. (2005) 'Recent changes to conservation of New Zealand's native biodiversity' *New Zealand Geographer*, 61, pp. 131–138. || Kakapo recovery programme (2009). http://www.kakaporecovery.org.nz

P. 34–35 | LAND FACT 12

6%: The percentage of the world's surface area covered in tropical rainforest.

75%: The total percentage of the earth's species of plants and animals that are found in tropical rainforest.

COMMENT: Species such as arthropods (e.g., insects and spiders) and flowering plants are found in high concentrations in tropical forests and a small area of tropical forest may have 100 or 200 species of trees of reasonable size. In addition to their importance for biodiversity, they have enormous economic value. The estimated value of tropical forests is thought to be as much as $0.8 trillion with ecotourism valued at around $3.0 trillion. In countries such as Costa Rica, Ecuador, the Philippines and Thailand, tourism brings in more foreign revenue than all the timber and timber products.

REFERENCE: Dogse, P. (1989) 'Sustainable Tropical Rainforest Management: Some Economic considerations'. UNESCO, Paris. || Gradwhol, J. and Greenberg, R. (1988) *Saving the Tropical Forests*, London: Earthscan. || Pearce, D. and Warford, J. (1993) *World without End: Economics, Environment and Sustainable Development*, Oxford: Oxford University Press. || Whitmore, T. (1998) *An Introduction to Tropical Rainforests*, Oxford: Oxford University Press.

P. 36–37 | LAND FACT 13

$1,175: Value per hectare per year of the Amazonian rainforest based on sustainable uses.

$110–150: Value per hectare per year if converted to pasture or other land use.

COMMENT: Looking specifically at the Amazonian tropical rainforest, various studies have estimated that its value was $1,175 per hectare per year. This value was made up of $549 ha/year for timber, resins, latex, food and recreation; $414 ha/year for its role in removing carbon from the atmosphere, preventing soil erosion, flood control; and $212 ha/year for possible future discoveries of medicine or other agricultural products as well as intangible benefits. The cost, therefore, of the deforestation of the tropical rainforest in Brazil alone was between 8–18% of their total GDP — highlighting the utter foolishness of such a waste of both natural and human resources. Estimates of the value of products obtained through commercial, non-sustainable uses of the rainforest. ranged from $150 per hectare per year for agricultural products to only $110 per hectare per year for pasture used for milk and calves — a fraction of the $1,175 per hectare per year value.

REFERENCE: Dogse, P. (1989) 'Sustainable Tropical Rainforest Management: Some Economic considerations',

UNESCO: Paris. || Gradwhol, J. and Greenberg, R. (1988) *Saving the Tropical Forests*, London: Earthscan. || Pearce, D. and Warford, J. (1993) *World without End: Economics, Environment and Sustainable Development*, Oxford: Oxford University Press. || Shone, B. and Caviglia-Harris, J. (2006) 'Quantifying and comparing the value of non-timber forest products in the Amazon' Ecological Economics, 58, pp. 249–267. || Torras, M. (2000) 'The total economic value of Amazonian deforest-ation, 1978–1993' *Ecological Economics*, 33, pp. 283–297.

P. 38–39 | LAND FACT 14

7,140: The area in square metres of an international football pitch.

1.5: The number of seconds you have to count to before another football pitch sized area of tropical rainforest is destroyed.

COMMENT: There is much debate on the actual rate of tropical forest destruction. In 2001, the FAO were reporting over 16.7 million hectares (1 hectare=10,000 square metres) with the highest rates of deforestation having occurred in Brazil (2.317mh), India (1.897mh) and Indonesia (1.687mh). Other sources estimated the loss to be only 9.6 mh, a difference of nearly 80%. However, even with the lowest possible estimate this still means you count to 3 seconds as opposed to 1.5 before a football field-sized area of tropical forest disappears.

REFERENCE: DeFries, R., Houghton, R., Hansen, M., *et al.* (2002) 'Carbon emissions from tropical deforestation and regrowth based on satellite observations for the 1980s and 90s' *Proceedings of the National Academy of Sciences*, 99 ,14256-14261.FAO — Food and Agriculture Organization (2001). *Global Forest Resources Assessment 2000: Main Report*. FAO Forestry Paper No. 140, Rome, Italy. || Houghton, J. (2004) *Global Warming: The Complete Briefing, third edition*, Cambridge: Cambridge University Press.

P. 40–41 | LAND FACT 15

40 million tonnes of greenhouse gases were emitted in 2007 from the mining of the Alberta Tar Sands. This is equivalent to the pollution of ten million cars.

142 million tonnes: The estimated amount of greenhouse gases which would be emitted by 2020 if the growth of the Tar Sands extraction is left unchecked.

32,000 square miles: The land area already approved by the Alberta government for exploration and development by the big oil companies.

COMMENT: The Tar Sands of Alberta hold the second largest oil reserve in the world, second only to Saudi Arabia. The Tar Sands are made up of layers of bitumen, sand and clay. Bitumen is a dense, viscous form of oil. Unlike conventional oil production, the process needed to extract and refine the oil from the Tar Sands produces not only far more greenhouse gases, but also a vast quantity of toxic waste water. The bitumen is extracted either by strip mining where acres of land are clearcut and the oil dug out, or by pumping vast quantities of steam or solvent into the sands to make the bitumen flow out more easily. It is estimated that over 3 million gallons of toxic waste water per day are seeping into the Athabasca water shed in Alberta. Scientists estimate that over 166 million migrating birds will die over the next several decades as a result of the mining of the Alberta Tar Sands. The pollution also has toxic effects on the human population with increased rates of cancer found in people living downstream from the mines.

REFERENCE: Hatch, C. and Price, M. (2008) 'The Most Destructive Project on Earth'. Environmental Defence. http://www.environmentaldefence.ca/reports/pdf/TarSands_TheReport.pdf || Tar Sands Watch, Polaris Institute, video 'Canada's Dirty Oil: Breaking our Addiction'. http://www.tarsandswatch.org

60 million: The estimated number of bison living in the Great Plains of America before the European colonisation of the 19th century.

1,000: The estimated number remaining after the great slaughters of 1870–1873 and 1880–1883

COMMENT: TThe Native American Indians by and large lived in a state of sustainability with the bison. The rapid European expansion of North America during the 19th century, however, brought European technology to what could be termed a pristine environment causing a massive drop in species numbers. It is estimated that it only took 200 years for 95% of the total North American forest to be destroyed in contrast to the 2,000 years that it took the Europeans to destroy a similar percentage

It was not just the bison and the forests that suffered in North America. Since the first homesteads of 1930, it is estimated that between 80–99% of the tall grass prairie has disappeared. This loss of habitat has led to estimates that up to 98% of the prairie dog population was lost with the consequent decline in species such as black-footed ferret, ferruginous hawk and the swift fox.

REFERENCE: Goudie, A. (2006) *The Human Impact on the Environment*, sixth edition, Oxford: Blackwell. || Marsh, R. (1984) 'Ground squirrels, prairie dogs and marmots as pests on rangeland', in *Proceedings of the conference for organisation and practice of vertebrate pest control*, ICI Plant protection division, Fernhurst, UK. || Samson, F. and Knopf, F. (1994) 'Prairie conservation in North America' *Bioscience*, 4.6, pp. 418–421.

1/3: The number of bird species that have become extinct in the mountain forests of San Antonio in the Colombian Andes.

80: The number of years in which this extinction has taken place.

COMMENT: Areas such as the Colombian Andes suffer biodiversity loss due to the fragmentation of their forests. Fragmentation occurs when parts of the forest are destroyed leaving isolated patches of forest. Evergreen tropical forests are particularly susceptible to species loss due to fragmentation as they contain a large number of indigenous species which are sparsely distributed. The main causes of such rapid declines in species numbers are due to the reduction in population sizes, the reduction of immigration rates, forest edge effects and the immigration of exotic species. This process of fragmentation represents as big a threat to the biodiversity of evergreen tropical forest as total clearances.

REFERENCE: Kattan, G., Alvarez-Lopez, H. and Giraldo, M. (1994) 'Forest fragmentation and bird extinctions: San Antonio eighty years later' *Conservation Biology*, 8, pp. 138–146. || Turner, I., (1996). 'Species loss in fragments of tropical forest: a review of the evidence' *Journal of Applied Ecology*, 33, pp. 200–209.

1/3: The amount of forest cover lost on Mt. Kilimanjaro in the last 70 years.

25%: The percentage of this forest cover that is irreplaceable upper montane and subalpine forest. The destruction has been caused by fires induced by changes in the climate.

COMMENT: Since 1900, the increasing loss of glaciers and snow cover on the summit of Mt. Kilimanjaro has been commented on. This loss is caused by both an increase in temperature and a reduction in precipitation. It has also had the effect of increasing the risk of fires on the upper slopes of the mountain as they become drier and hotter. This change in climate has not only had a negative impact on the species diversity of these forests but also has had

an impact on the water balance of the area. As the montane forest is lost, fog interception is also lost meaning that the moisture that would have been captured from the air by the trees is not available to provide water for the population living on the mountain.

REFERENCE: Altmann, J., Alberts, S., Altmann, S., *et al.* (2002) 'Dramatic change in local climate patterns in the Amboseli Basin, Kenya' *African Journal of Ecology*, 40, pp. 248–251. || Hemp, A. (2005) 'Climate change-driven forest fires marginalize the ice cap wasting on Mt. Kilimanjaro' *Global Change Biology*, 11, pp.1013–1023. || Hemp, A. (2005) 'The impact of fire on diversity, structure and composition of Mt. Kilimanjaro's vegetation' in E. Spehn, M. Liberman and C. Korner (eds.) *Land Use Changes and Mountain Biodiversity*, Boca Raton, FL: CRC Press LLC. || Hemp, A. (2009) 'Climate change and its impact on the forest of Kilimanjaro' *Journal of Ecology*, 47,pp. 3–10. || Meyer, H. (1900) *Der Kilimandscharo. Reisen und Studien.* Reimer,Berlin.

25%: The number of prescribed drugs that have their base in products found in tropical forests.

$500 billion: The total value of medical products based on products found in tropical forests.

COMMENT: 25% of all prescribed drugs have their base in products found in tropical forests and in poorer countries this rises to 70% . One of the most famous medical discoveries made in tropical forests was the discovery of quinine derived from the bark of the cinchona tree and used to treat malaria. Since its discovery in the 17th century, it has been adapted and used to treat many forms of cancer such as leukaemia and has immense economic value. Quinine has a bitter taste and, in addition to its medicinal uses, it is the ingredient in both bitter lemon and the tonic used in gin and tonics.

REFERENCE: Pearce, D. and Moran, D. (2004) *The Economic Value of Biodiversity*, London: Earthscan. || Principe, P. (1987) 'The Economic Value of Biological Diversity among Medicinal Plants', OECD: Paris. || Principe, P. (1991) 'Valuing the biodiversity of medicinal plants' in O. Akerete, V. Heywood and H. Synge *The Conservation of Medicinal Plants*. Cambridge: Cambridge University Press.

4: The number of tonnes of top soil for each woman, man and child on the planet that is lost every year due to soil erosion.

10 million: The number of hectares of arable land per year that is being lost due to soil erosion.

COMMENT: Soil erosion is a major environmental threat to the continued productive capacity of agriculture. With the world's population set to reach nine billion by 2020, the demand for food is increasing at a time when the capacity of the planet to feed the growing population is declining. Furthermore, the soil erosion caused by agriculture far exceeds even the impact of Pleistocene continental glaciers or the current impact of alpine erosion by glacial and fluvial processes. Since 1961, although the area of cropland has increased by 11%, the global population has approximately doubled. The net effect of these changes is that per capita cropland area has decreased by 44% even before factoring in the inevitable loss of productivity caused by soil erosion. The world, therefore, faces a looming crisis of population growth and food production decline.

REFERENCE: Pimentel, D., Harvey, C., Resosudarmo, P., *et al.* (1995) 'Environmental and Economic Costs of Soil Erosion and Conservation Benefits' *Science*, 267, pp. 1117–1123. || Wilkinson, B. and McElroy, B. (2007) 'The impact of humans on continental erosion and sedimentation' *Geological Society of America Bulletin*, 119.1, pp. 140–156.

80: The amount of nitrogen (in Teragrams Nitrogen/ Year) emitted into the atmosphere in the form of ammonia (NH_3) caused by human activity. [1 teragram = 1,000,000,000 kilograms].

20: The amount of nitrogen (in Tg N/Yr) emitted into the atmosphere through burning fossil fuels.

COMMENT: During the 1980s and 1990s concern over the acidification of lakes and soils was centred on the combustion of fossil fuels and the subsequent release of sulphur dioxide and nitrogen dioxide, but in fact more acidification of soils on a global scale occurs from the production of fertilizers. In fact NH_3 has twice the potential to acidify soils as nitrogen dioxide. Indeed only 40% of the anthropogenic nitrogen production comes from developed countries, 60% comes from the developing world. With populations increasing both in China and India and the use of fertilizers also increasing, the acidification of soils is set to rise uncontrollably.

REFERENCE: Galloway, J., Schlesinger, W., Levy II, H., *et al.* (1995) 'Nitrogen fixation: Anthropogenic enhancement— an environmental response' *Global Biogeochemical Cycles*, 9, pp. 235–252. || Vitousek, P., Mooney, H., Lubchenco, J. *et al.* (1997) 'Human domination of earth's ecosystems' *Science*, 277, pp. 494–499.

30%: The amount of applied nitrogen fertilizer that is taken up by crops.

67% : The reduction in the nitrogen efficiency of crops between 1960 and 2000.

COMMENT: Intensive high-yield agriculture is dependent on the addition of fertilizers, especially industrially produced NH_4 and NO_3. However, a significant amount of the applied nitrogen and a smaller portion of the applied phosphorus are lost from agricultural fields and may end up as runoff into rivers and oceans. Since pre-industrial times, the amount of nitrogen from rivers running into the North Atlantic has increased by 2 to 20-fold. This pollution harms off-site ecosystems, water quality and aquatic ecosystems, and contributes to changes in atmospheric composition. Perhaps crucially, nitrogen inputs to agricultural systems contribute to emissions of the greenhouse gas nitrous oxide. Indeed it has been suggested that rice paddy agriculture and livestock production are the most important anthropogenic sources of the greenhouse gas methane.

REFERENCE: Prather, M. in IPCC (2001) *Climate Change 2001: The Scientific Basis, Contribution of Working Group I to the Third Assessment Report of the Intergovernmental Panel on Climate Change* [J. T. Houghton *et al* (eds..)] Cambridge: Cambridge University Press. http://www.csun. edu/~hmc60533/csun_630E_S2004/climate change/climate_ change_2001_tech_summary.pdf || Tilman, T., Cassman, K., Matson, P., *et al.* (2002) 'Agricultural sustainability and intensive production practices' *Nature*, 418, pp. 671–677.

30%: The decline in common farmland birds in Europe over the period 1978 to 2002.

100%: The increase in the use of pesticides over the same period.

COMMENT: The use of pesticides has alarming consequences not only for the agricultural ecosystem, but also for the surrounding ecosystems — including humans. DDT (now banned) was successful in reducing the pests but could not be broken down by the animals. It was therefore passed progressively up through the food chain until it killed high order predators and caused birds to lay eggs with shells too thin to support life. Other pesticides reduce the insects that farmland birds rely on for food, while changes in agricultural practices either reduce the habitat of birds such as the corncrake or reduce the number of seeds from wild flowers through the use of herbicides.

REFERENCE: Gregory, R., Vorisek, P., van Strien, A., *et al.* (2003) 'From bird monitoring to policy-relevant indicators: A report to the European Topic Centre on Nature Protection and Biodiversity', The Royal Society for the Protection of Birds, Bedfordshire, UK. || Tilman, D., Fargione, J., Wolff, B., *et al.* (2001)' Forecasting agriculturally driven global environmental change' *Science*, 292, pp. 281–284.

25th September: The day when the world goes into ecological debt in terms of its resource consumption and waste disposal.

25th November: The equivalent debt day back in 1995 highlighting the fact that we are over-using our resources at an ever increasing pace.

COMMENT: There has been considerable work on what some people have termed a neo-Malthusian approach to resource consumption. According to the New Economics Foundation, the unwavering pursuit of economic growth — embodied in the overwhelming focus on Gross Domestic Product (GDP) — has left over a billion people in dire poverty. It has not notably improved the well-being of those who were already rich, nor even provided us with economic stability. Instead it has brought us straight to the cliff edge of rapidly diminishing natural resources and unpredictable climate change. Their analysis confirms that the countries where people enjoy the happiest and healthiest lives are carrying out unsustainable ecological practices — using more resources than the earth can provide on a sustainable basis. According to the NEF, we are building up such an ecological debt that the world ecosystems will eventually catastrophically collapse.

REFERENCE: New Economic Foundation (2009) 'The Happy Planet Index 2.0: why good lives don't have to cost the earth'. http://www.happyplanetindex.org

3: Out of the 9 critical natural boundaries for earth the number that have already been passed. These three are climate change, rate of biodiversity loss and interference in the nitrogen cycle.

4: The number of the remaining 6 critical natural boundaries that are in imminent danger of being passed: global freshwater use, change in land use, ocean acidification, and interference with the global phosphorus cycle.

COMMENT: Scientists have suggested that since the Industrial Revolution a new era has arisen which they termed the Anthropocene. In this era human actions have become the main driver of global environmental change. They then hypothesized that human activities such as our growing reliance on fossil fuels and industrialized forms of agriculture could push the earth system outside the stable environmental state of the preceding era, the Holocene, with consequences that are detrimental or even catastrophic for large parts of the world. It is now believed that 3 out of the 9 natural boundaries have already been passed and 4 more are in imminent danger. The remaining two natural boundaries which are currently not in danger are stratospheric ozone depletion and chemical pollution.

REFERENCE: Rockström, J., Steffen,W., Noone, K., *et al.* (2009) 'Planetary boundaries: exploring the safe operating space for humanity' *Ecology and Society*, 14.2:32. www.ecologyandsociety.org/vol14/iss2/art32/

P. 62–63 | LAND FACT 26
9.4: The number of global hectares per capita that the United States of America needs to maintain its present consumption pattern.

4: The number of earths needed to meet this demand.

COMMENT: Another way to measure whether people are exceeding the capacity of the planet to provide sustainable resources is by looking at our ecological footprint. An ecological footprint is an estimate of the use of resources by a country in one year. To achieve one-planet living, a country must keep its ecological footprint below the level that corresponds to its fair share given the world's current biocapacity and population. This number is 2.1 global hectares (or gha) per person. The United States of America has increased its footprint from 8.2 to 9.4 since 1990. The poorest countries, such as Malawi, Haiti and Bangladesh, have the smallest per capita footprints. The ecological footprint does not take into account environmental damage created by a country such as air pollution or deforestation. The figure of 2.1 allows for humans only without providing any capacity for other species. We should actually be using no more than 70–80% of the earth's capacity to take this into account.

REFERENCE: New Economic Foundation (2009) 'The Happy Planet Index 2.0: why good lives don't have to cost the earth'. http://www.happyplanetindex.org || Price, A. (2009) *Slow-tech: Manifesto for an Over-wound World*, London: Atlantic Books.

P. 64–65 | LAND FACT 27
1: The number of species (humans) using 86.6% of the world's biocapacity

1.7 million: The number of known species using the remaining 13.4% of the world's biocapacity

COMMENT: Ecofootprint 1.0 did not take into consideration that the world's biocapacity (i.e. resources) should be allocated to all the species living on earth, not just humans. Ecofootprint 2.0 adjusted this number to allocate 13.4% of the world's biocapacity to these other species . This effectively meant that the original allocation underestimated the extent to which humans are living beyond the capability of the earth to support us. Ecofootprint 2.0, therefore, represents a truer picture of our unsustainable lifestyle.

REFERENCE: Chambers, N., Simmons, C. and Wackernagel, M. (2000) *Sharing Nature's Interest: Using Ecological Footprints as an Indicator of Sustainability*, London: Earthscan. || Venetoulis, J. and Talberth, J. (2006) 'Ecological Footprint of Nations 2005 UPDATE – Sustainability Indicators Program', Redefining Progress www.redefiningprogress.org

Chapter Two: The Oceans
P. 68–69 | OCEANS FACT 1
8.1: The average pH of ocean surface water in 2009. In 1800, the figure was pH 8.2. The oceans have slowly become less alkaline as they have absorbed the excess carbon dioxide released by burning fossil fuels.

7.8: The estimated pH of the oceans by 2070 if atmospheric carbon dioxide levels continue to rise to 840 parts per million. More acidic oceans will threaten the life-cycles of coral and many other marine organisms.

COMMENT: Changes in ocean acidity and alkalinity may seem small; however they are likely to be significant for marine organisms which secrete a calcium carbonate shell or exoskeleton. These organisms include corals, molluscs and crustaceans. More acidic ocean water may mean the process of making calcium carbonate structures will be more difficult and the structures weaker and more prone to being dissolved.

REFERENCE: The Royal Society, 'Ocean acidification due to increasing atmospheric carbon dioxide', Policy document 12/05, June 2005, http://www.royalsociety.org/ocean-acidification-due-to-increasing-atmospheric-carbon-dioxide/

P. 70–71 | OCEANS FACT 2
1 million tonnes per hour: The amount of anthropogenic carbon dioxide being absorbed by the oceans

5,000–10,000 years: The timescale for natural changes to ocean acidity of a similar magnitude to that caused by humans in only 50–100 years.

COMMENT: Ocean acidification may seem like a very minor change, but like many of the changes associated with global warming, the speed of the change is worrying. Changes on timescales of hundreds and thousands of years are slow enough to allow nature to adapt – at a very simple level to move to new locations or even evolve to cope. Very rapid change makes this adaptation impossible and as a result raises the spectre of ecosystem degradation and extinction.

REFERENCE: Brewer, P.G. and Barry, J. (2008) 'The Other CO2 problem' *Scientific American, Earth 3.0 special issue*, 18.4, pp. 22–23.

P. 72–73 | OCEANS FACT 3
18 cm: The global rise in sea level between 1900 and 2000, based on a rise of 1.8mm per year.

90–130 cm: The estimated sea level rise between 2000 and 2100. Sea levels will rise due to thermal expansion as the oceans continue to warm, plus additional water from melting ice caps, glaciers and permafrost.

COMMENT: Sea level change is notoriously difficult to measure. One reason for this is that across much of the world land levels are slowly changing, often as a result of tectonic movement. The figure of 1.8 mm / year during the 20th century is generally accepted as reliable. Future projections of sea level vary widely. In 2007 the Inter-Governmental Panel on Climate Change (IPCC) estimated an 18–59 cm rise by 2099 depending on exact future emissions, energy sources, world population and economic growth scenarios. These estimates excluded the contribution that might come from the melting of the Greenland and Antarctic ice caps. More recent research, factoring in some melting of ice caps, suggest sea level rise by 2100 of around 1m.

REFERENCE: Bindoff, N.L., Willebrand, J., Artale, V., et al. (2007) 'Observations: Oceanic Climate Change and Sea Level' in *Climate Change 2007: The Physical Science Basis, Contribution of Working Group I to the Fourth Assessment Report of the Intergovernmental Panel on Climate Change* Cambridge and New York: Cambridge University Press. http://www.ipcc.ch || Grinsted, A., Moore, J.C. and Jevrejeva, S (2009) 'Reconstructing sea level from palaeo and projected temperatures 200 to 2100 AD', *Climate Dynamics*, 34.4, pp. 461–472.

P. 74–75 | OCEANS FACT 4
1.5–2.0°C: The rise in sea temperatures that results in coral reef bleaching events, leading to the degradation and death of the reefs.

2.8°C: The average projected increase in global temperatures by 2099.

COMMENT: Coral is highly sensitive to ocean temperature change. As temperatures rise the zooanthella (algae) which provide the coral animal with much of its energy are driven out of the coral skeleton. Zooanthella are vital to coral health; without them coral quickly dies leaving only the bleached white corraline skeleton. In 1998 ocean temperatures rose across much of the world in response to an El Nino event. This event killed or damaged 20% of reefs worldwide – perhaps a foretaste of the future.

REFERENCE: Australian Government, Great Barrier Reef Marine Park, 'Climate change', http://www.gbrmpa.gov.au/corp_site/key_issues/climate_change/climate_change_

|| Meehl, G.A., Stocker, T.F., Collins, W.D., *et al.* (2007)
'Global Climate Projections' in *Climate Change 2007: The Physical Science Basis, Contribution of Working Group I to the Fourth Assessment Report of the Intergovernmental Panel on Climate Change*, Cambridge and New York: Cambridge University Press. http://www.ipcc.ch/pdf/assessment-report/ar4/wg1/ar4-wg1-chapter10.pdf

P. 76–77 | OCEANS FACT 5

85%: The percentage of the world's major river deltas which are sinking according to a study published in 2009

50%: The projected increase in the area of delta land that will be flooded by 2050 due to a combination of sinking land and rising sea levels, putting up to 500 million people at risk.

COMMENT: River deltas have very high human population densities because the land is usually very fertile and flat and water is plentiful. This makes deltas ideal for farming. Many of the world's major cities are located on or close to deltas. They are ideal locations for trade, sitting where major rivers meet the ocean. It has been found, however, that 85% of the 33 major deltas studied were found to be sinking. While deltas naturally sink under their own weight, this sinking is usually balanced by new sediment deposition as rivers carry sediment from further upstream to the delta. However dam construction has in many cases blocked the transport of this sediment, robbing deltas of their supply. Extraction of groundwater from deltas, for irrigation, urban areas and industry is also a major cause of sinking.

REFERENCE: Syvitski, J.P.M., Kettner, A.J., Overeem, I., *et al.* (2009) 'Sinking deltas due to human activities', *Nature Geoscience 2*, pp. 681–686.

P. 78–79 | OCEANS FACT 6

-1 to -3°C: The cooling the Northwest of Europe could experience if the warm ocean currents from the Gulf Stream and North Atlantic Drift collapsed due to global warming. The Norwegian coast might experience a cooling of up to -12°C.

COMMENT: Northwest Europe is warmed by ocean currents which form part of the global thermohaline circulation. The UK and Scandinavia are significantly warmer than their latitudes might suggest due to the effect of warm ocean currents. Some climate models suggest global warming might disrupt or even shutdown some ocean currents which are critical to maintaining earth's climate patterns. While this scenario is less likely than other changes that might result from a warming world, it is certainly possible. Rapid melting of the Greenland ice sheet or significant changes to rainfall patterns could, in theory, cause warm, low density freshwater to flow into the western North Atlantic disrupting current temperatures and salinity patterns making such a scenario more likely.

REFERENCE: Osborn, T. and Kleinen, T. (2008) 'The thermohaline circulation', Climatic Research Unit Information Sheet no. 7 http://www.cru.uea.ac.uk/cru/info/thc/thc.pdf || Vellinga, M. and Wood, R.A. (2007) 'Impacts of thermohaline circulation shutdown in the twenty-first century', *Climatic Change*, 91.1–2, pp. 43–63.

P. 80–81 | OCEANS FACT 7

51: Small Island Developing States (SIDS) are especially vulnerable to climate change and sea level rise. Many are low lying, isolated and depend heavily on the oceans for resources.

0.02%: The proportion of global greenhouse gas emissions from the 51 SIDS.

COMMENT: The 51 SIDS are on the front line of climate change. Many low lying island chains are at risk of inundation caused by rising sea levels leading to the loss of a significant amount of land. The Maldives, Kiribati, Vanuatu and Tuvalu are among the most at risk. In addition, rising sea levels turn fresh groundwater resources saline, reducing water supply and contaminating the little farmland many SIDS possess. Increased storm and cyclone activity could cause more frequent flooding and storm damage. Many SIDS rely on coral reefs for fishing and tourism. A rise in sea temperature could cause the destruction of the coral reefs thereby directly reducing income.

REFERENCE: Roper, T. (2007) 'The IPCC Report and its implications for the Pacific', Pacific Power, 15.3, pp. 24–25, 30. http://www.gseii.org/PDF/roper-ipcc-report-implications.pdf || Mimura, N., L., Nurse, R.F., McLean, J., *et al.* (2007) 'Small islands' in *Climate Change 2007: Impacts, Adaptation and Vulnerability. Contribution of Working Group II to the Fourth Assessment Report of the Intergovernmental Panel on Climate Change*, M.L. Parry, O.F. Canziani, J.P. Palutikof, P.J. *et al.* (eds.), Cambridge: Cambridge University Press, pp. 687–716.

P. 82–83 | OCEANS FACT 8

865,000: One estimate of the combined population of minke, humpback and fin whales in the North Atlantic prior to the onset of commercial whaling.

215,000: Estimate of the combined population of these whales since the 1990s.

COMMENT: It has been argued that past estimates of 'natural' whale populations are likely to be inaccurate. For instance the pre-whaling population of humpback whales in the North Atlantic used by the International Whaling Commission (IWC) is 20,000, but the 2003 research suggests that the actual historic population was closer to 240,000. It is considered 'safe' by the IWC to resume commercial whaling when a population is at 54% of their historic levels. There are thought to be around 10,000 humpback whales in the North Atlantic today almost reaching the 54% guideline if the IWC number of 20,000 is used. If the historic number was, in fact, 240,000 the current population would need to be over 125,000 before whaling would be resumed. Fundamentally it is very difficult for humans to know what the population of any marine species was prior to the onset of human activity.

REFERENCE: Roman, J. and Palumbi, S.R. (2003) 'Whales Before Whaling in the North Atlantic' *Science*, 301.5632, pp. 508 – 510. || International Whaling Commission, Whale population estimates. http://www.iwcoffice.org/conservation/estimate.htm#table

P. 84–85 | OCEANS FACT 9

One quarter: of all marine species inhabit coral reefs, making them the most biodiverse oceanic environment.

One third: of all reef building coral species face extinction due to a range of threats including climate change, coastal development and unsustainable fishing.

COMMENT: In 2008, the status of 845 species of reef building coral was assessed by scientists using the International Union for Conservation of Nature Red List Criteria. Of the 704 species for which enough data existed to make an assessment, 32.8% were classified as having an elevated risk of extinction. This level of threat is especially worrying given that new species of coral continue to be discovered, for instance 3 new species were discovered in the Galapagos Islands in 2009 and around 150 were discovered in the Great Barrier Reef in 2008. There is a significant danger that many species of coral and other ocean dwelling animals and plants will become extinct before they have been discovered and described.

REFERENCE: Carpenter, K. E., *et al.* (2008) 'One-Third of Reef-Building Corals Face Elevated Extinction Risk from Climate Change and Local Impacts' *Science*, 321. 5888, pp. 560 – 563.

35 million tonnes: The global wild fish catch in 1960 when world population stood at 3 billion.

92 million tonnes: The global wild fish catch in 2006 with a world population continuing to rise past 6.5 billion.

COMMENT: The global wild fish catch appears to have reached a plateau, with some scientists suggesting 'peak wild fish' has been reached. Global production has been at around 90–95 million tonnes since the mid 1990s. Total fish production continues to rise due to the expansion of farmed fish (aquaculture). Unfortunately, this frequently comes at the expense of coastal mangrove swamps which are cut down and converted into fish farming ponds. In 2008 according to the UN Food and Agriculture Organisation, 80% of global wild fisheries were already either fully or over exploited.

REFERENCE: FAO Fisheries and Aquaculture Department (2009) 'The State of World Fisheries and Aquaculture 2008'. Food and Agriculture Organization of the United Nations. Rome. http://www.fao.org/fishery/sofia/en

11 of the 18 species of penguin show evidence of declining populations. A further 7 penguin species are classified as vulnerable and 4 as endangered.

COMMENT: Penguins are an iconic conservation species. Their oceanic and terrestrial habitats are threatened by climate change, fishing and coastal development. The International Union for the Conservation of Nature's 2009 Red List has identified 11 species of penguin as at risk of extinction (in the risk categories of critically endangered, endangered or vulnerable). Penguins rely on a diet of fish, so any changes to the numbers of organisms further down the food chain will inevitably affect the availability of food for penguins.

REFERENCE: IUCN (2009) 'IUCN Red List of Threatened Species', http://www.iucnredlist.org

250 parts per million: The level of PCB contamination found in some northern Pacific killer whales

1 part per million: The level of PCB contamination found in most humans.

COMMENT: PCBs (Polychlorinated biphenyls) are chemicals once widely used in electronics and as coolants. PCB production was banned in the United States in 1979 and restricted globally by the 2001 Stockholm Convention on Persistent Organic Pollutants. PCBs are very stable chemicals which do not easily degrade in the environment so despite bans, they continue to cause problems. Chemical waste entering the oceans via rivers gradually works it way into and up the food chain concentrating in the bodies of top predators such as killer whales and polar bears. PCBs lodge in the fatty tissues of predators and cannot be broken down so, for example, as a killer whale eats more salmon contaminated by PCBs the level of the chemical continues to build up in its body. Some male killer whales can have PCB levels of 250 parts per million, the highest concentration recorded in a mammal. PCBs increase vulnerability to infectious disease, impair reproduction and impede normal growth and development. Estimates suggest PCBs could still be having an impact on marine mammals in the 2050s due to their persistence in the environment.

REFERENCE: Hickie, B.E., Ross, P.S., Macdonald, R.W. *et al.* (2007) 'Killer whales (Orcinus orca) face protracted health risks associated with lifetime exposure to PCBs' *Environmental Science and Technology* 41.18 pp. 6613–6619. || Cone, M. (2009) 'Poisoned killer whales? Blame salmon.' *Scientific American*. January 20, 2009. http://www.scientificamerican.com/article.cfm?id=salmon-to-blame-for-poisoned-killer-whales || Anderson, G. (2003) 'Killer

whales: outlook' in *Marine Science*. http://www.marinebio.net/marinescience/05nekton/KWoutlook.htm

40%: Of the global marine fish catch is bycatch, much of which is simply discarded overboard.

COMMENT: In commercial fishing, the nets used to catch a particular type of fish (target catch) also capture unintended species (non-target catch or bycatch). Estimates of bycatch vary widely, from around 10% for some well managed and very selective fisheries up to 90% for poorly managed shrimp trawling. Reports for the UN Food and Agriculture Organisation estimated global bycatch in the region of 27 million tonnes in 1994, but only 7 million in 2004. The difference is accounted for by improving fishing methods and differences in the definition of bycatch. Bycatch is defined as non-target species which are caught and then discarded at sea, landed but not sold (and therefore wasted) plus species which are caught but are not managed. This last group recognises that catching unmanaged species is inherently unsustainable because the 'safe' quantity is not known.

REFERENCE: Davies, R.W.D., Cripps, S.J., Nickson, A., *et al.* (2009) 'Defining and estimating global marine fisheries bycatch' *Marine Policy*, 33:4, pp. 661–672.

2048: One estimate of the year global marine fish stocks will have collapsed by 90%, effectively meaning the end of the oceans as a useful source of wild food.

9 billion : The number of mouths to feed in 2048, at current rates of global population growth.

COMMENT: The scientists who published the 2048 estimate in 2006 calculated that by 2003, 29% of all commercially caught marine species had collapsed. 'Collapse' is defined as the size of the catch having fallen to 90% of the maximum catch in the past. Overfishing of cod in the Grand Banks off Newfoundland, led to the closure of the fishery in 1992 and the loss of around 40,000 jobs in the fishing and fish processing industries. Despite efforts to promote the recovery of cod, the fishery has not reopened and stocks of mature cod have never returned. A similar situation repeated globally would have serious consequences for food supply in many nations.

REFERENCE: Worm, B., Barbier, E.B., Beaumont, N., *et al.* (2006) 'Impact of Biodiversity Loss on Ocean Ecosystem Services', *Science*, 314:5800, pp. 787–790.

50%: The percentage of the world's wild fish which are caught in an area covering only 7.5% of the world's oceans, meaning that food resources from the oceans are highly geographically concentrated.

COMMENT: It would not be unreasonable to think that the world's oceans are teeming with life just waiting to be caught and cooked. In fact, this is only true in select locations such as the shallow seas on the continental shelf. The vast expanses of deep ocean further from land actually yield very little in the way of marine food resources. Unfortunately the 7.5% of the world's oceans which are the richest source of food are also those areas closest to land and large cities, both of which increasingly pollute the sea. These small areas are increasingly over-fished, threatening the long-term sustainability of the oceans as a source of food for humans.

REFERENCE: Nellemann, C., Hain, S. and Alder, J. (eds.) (2008) 'In Dead Water — Merging of climate change with pollution, over-harvest, and infestations in the world's fishing grounds'. United Nations Environment Programme, GRID-Arendal, Norway. http://www.unep.org/pdf/InDeadWater_LR.pdf

9.9 kilograms: The worldwide average annual consumption of fish per person in the 1960s.

16.7 kilograms: The worldwide average annual consumption of fish in 2006 per person. About 1 billion people rely almost solely on fish as a source of protein.

COMMENT: Fish consumption per person has almost doubled since the 1960s. At the same time the global population has also more than doubled. These two facts serve to illustrate the pressure on the oceans as a source of food. For the large number of people who depend almost entirely on the oceans for protein, the continued degradation of fish stocks is likely to have a severe adverse effect on their diet and health. Global trends suggest that wild fish catches have reached a plateau and may even be declining. In the last twenty years aquaculture (fish farming) has grown significantly, allowing the total fish supply to rise even though wild fish capture has not increased. In 1970 aquaculture was less than 4% of global fish production, but by 2006 this had risen to 50 million tonnes or 36%.

REFERENCE: FAO Fisheries and Aquaculture Department (2009) 'The State of World Fisheries and Aquaculture 2008'. Food and Agriculture Organization of the United Nations, Rome. http://www.fao.org/fishery/sofia/en

1%: The percentage of the world's fleet of 3.5 million fishing boats which are large, factory fishing vessels.

50%: The percentage of the world's annual marine fish catch of 80–90 million tonnes which is caught by large, factory fishing vessels.

COMMENT: The vast majority of the world's fishing boats are small, operated by one or two people and operate close to shore. Large industrial 'factory' fishing boats are rare. The Food and Agriculture Organisation of the UN estimates the worldwide number of vessels of over 100 tonnes to be around 25,000. These ships may stay at sea for several months because they are able process and freeze fish on board. They are responsible for the majority of the world's fish catch. Some estimates suggest that there are also 1,000–1,500 'pirate' fishing vessels worldwide that operate illegally and are unregulated. These vessels represent a serious threat to sustainable fisheries.

REFERENCE: Greenpeace Factsheet (2008) 'Oceans', http://www.greenpeace.org/raw/content/australia/resources/fact-sheets/overfishing/oceans-under-threat.pdf || FAO Fisheries and Aquaculture Department (2004) 'The State of World Fisheries and Aquaculture 2004'. Food and Agriculture Organization of the United Nations. Rome. http://www.fao.org/fishery/sofia/en

200,000 loggerhead sea turtles and 50,000 leatherbacks caught up in fishing gear worldwide. Populations of both species have fallen by 80–90% in the past decade.

COMMENT: Marine turtles are extremely long-lived animals. A loggerhead reaches reproductive age at about 35 years old and can expect to live for over 50 years. Both turtle species are large, up to two metres long, and are therefore vulnerable to being entangled in nets with a small mesh size. Because turtles are air breathing, they must surface to breathe — if they become entangled in fishing gear they can drown.

REFERENCE: UNEP (2006) 'Ecosystems and Biodiversity in Deep Waters and High Seas' UNEP Regional Seas Report and Studies No. 178, UNEP/IUCN, Switzerland, 2006 http://www.icriforum.org/secretariat/cold/IUCN_Report_16June06.pdf

20%: The percentage of coastal mangrove forests which have been destroyed worldwide since 1980, amounting to a loss of 3.6 million hectares of mangrove.

Mangroves are critically endangered or approaching extinction in 26 of the 120 countries where mangrove forests are found.

COMMENT: Mangrove forests are a unique ecosystem straddling the edge of both the oceans and the land. They provide a natural coastal buffer against storms and erosion as well as acting as protective nurseries for young fish. Mangroves face many threats.
The two most serious are deforestation and drainage for farmland or conversion to aquaculture ponds to farm shrimp. Deforestation for fuelwood and urbanisation are also important threats. Deforestation leaves the coastline vulnerable to flooding and, in the long-term, depletes fish stocks. It also means that local people lose a sustainable source of fuelwood.

REFERENCE: Food and Agriculture Organization of the United Nations (2007) 'The world's mangroves 1980–2005', FAO Forestry Paper 153, Rome. ftp://ftp.fao.org/docrep/fao/010/a1427e/a1427e00.pdf

20 million tonnes: The amount of wild marine fish used every year to produce fish food for farmed fish.

5 kilograms: The number of kilograms of wild marine fish needed to produce food to feed 1 kilogram of farmed salmon to maturity.

COMMENT: Strange as it may seem, much of the food used in aquaculture is actually produced using fish protein from wild caught fish such as sardines, anchovies and sand eels. In 2006 around 20 million tonnes of wild fish were used in this way. Globally, for each kilogram of farmed fish 0.6 kilograms of wild fish are used. Some species require a much richer diet, for example 1 kg of farmed salmon eats 5 kg of wild fish based fishmeal. It may seem reasonable to use fish which humans do not eat as food for aquaculture, but overfishing of these fish stocks will have a knock-on effect for species in the food web which humans do consume directly. The UN FAO has reported that fishmeal and fish-oil sources could be depleted by 2030, putting pressure on aquaculture to find new food protein sources.

REFERENCE: Naylor, R. L., Hardy., R.W., Bureau, D.P., *et al.* (2009) 'Feeding aquaculture in an era of finite resources' in *Proceedings of the National Academy of Scientists*, 106.34. http://www.pnas.org/content/106/36/15103.full.pdf+html

100 million tonnes of floating rubbish is believed to be trapped in the Great Pacific Garbage Patch, much of it plastic waste. Vast circular ocean currents, or gyres, sweep rubbish into continental scale waste patches.

18,000: Pieces of plastic trash in every square kilometre of ocean.

COMMENT: Virtually all plastic rubbish is non-biodegradable. Plastic waste in the oceans can originate from shipping, such as discarded or lost fishing equipment or rubbish thrown over-board. 80% of oceanic plastic waste is thought to originate from rivers draining into the oceans. In 1997 Charles Moore discovered the Great Pacific Garbage Patch, the resting place of millions of tonnes of plastic and other waste which will continue to pollute the oceans for centuries.

REFERENCE: Tamanaha, M. and Moore, C., 'Plastics Are Forever', Algalita Marine Research Foundation. http://www.algalita.org/pdf/plastics%20are%20forever%20english.pdf || UNEP (2006) 'Ecosystems and Biodiversity in Deep Waters and High Seas' *UNEP Regional Seas Report and Studies No. 178*, UNEP/IUCN, Switzerland. http://www.icriforum.org/secretariat/cold/IUCN_Report_16June06.pdf

P. 110–111 | OCEANS FACT 22

1.3 million tonnes: The amount of oil discharged into the sea, worldwide, each year: Equivalent to 3 of the world's largest super-tankers discharging their entire cargo. (This figure doesn't take into account the 4.9 million barrels into the sea by the Gulf of Mexico oil spill of 2010).

COMMENT: The world's largest super-tankers, called Ultra Large Crude Carriers, can carry 430,000 tonnes of crude oil. The 2010 Gulf of Mexico oil-rig spill was, however, the worst ever with over 4.9 million barrels, far more than the estimated 37,000 tonnes spilled from the infamous *Exxon Valdez* in 1989. Most marine oil pollution does not originate from major spills, but rather from natural seepage from rocks, numerous small spills from oil production rigs, ships and pipelines, and spills on land that are carried into the sea via rivers. Oil coats the bodies of marine birds and seals becoming stickier over time and harder to remove. Oil degrades the natural insulating properties of feathers and fur causing hypothermia, which can be fatal. In addition marine life may ingest oil, which is toxic, leading to poisoning, reducing breeding and increasing their vulnerability to disease and infection.

REFERENCE: Committee on Oil in the Sea: inputs, fates and effects and National Research Council (2003) *Oil in the Sea III: inputs, fates and effects*' Washington, D.C.: National Academies Press. http://www.nap.edu/catalog.php?record_id=10388#toc

P. 112–113 | OCEANS FACT 23

14 %: Decline in the global marine Living Planet Index 1970–2005.

40%: of the oceans are severely affected by human activities, but less than 1% of the oceans are classified as protected areas.

COMMENT: The *Living Planet Index*, published by the World Wildlife Fund, is an index of the state of the earth's biodiversity. It tracks the health of populations of animal species. The marine species index tracks 1,175 populations of 341 marine species. This index has shown a significant decline, especially since 1990. Much of the decline is a result of overfishing, but other stressors such as pollution and climate change are also having an impact on marine health.

REFERENCE: C. Hails (editor in chief) 'Living Planet Report 2008', World Wide Fund For Nature (formerly World Wildlife Fund) WWF INTERNATIONAL, Gland, Switzerland. http://assets.panda.org/downloads/living_planet_report_2008.pdf

P. 114–115 | OCEANS FACT 24

33%: Plastic bags and other plastic debris account for one third of all deaths of leatherback sea turtles.

COMMENT: Leatherback turtles, currently on the critically endangered list, feed on jellyfish. It has often been thought that leatherbacks may mistake floating plastic bags for jellyfish. In a recent report, it was shown that plastic was present in the gut of 37% of leatherback turtles from a sample of over 350 autopsies since 1968. Plastic blocks the digestive tract of turtles, leading to starvation and eventual death.

REFERENCE: N. Mrosovsky, Ryan, G., James, M. (2009) 'Leatherback turtles: The menace of plastic' *Marine Pollution Bulletin*, 58..2, pp. 287–289.

P. 116–117 | OCEANS FACT 25

405: The number of dead zones reported in the world's oceans in 2008. These are areas so depleted of oxygen they can no longer support life.

1–2% of the ocean floor is biologically dead or close to dying.

COMMENT: Dead zones are hypoxic, that is the amount of dissolved oxygen is so low that life cannot be supported. Most dead zones are just offshore from major cities. The largest concentrations are along the east and south coasts of the USA, northern Europe and parts of Asia. Dead zones result from pollution, including fertilizer runoff from fields, sewage and other urban waste. High levels of nitrate and phosphates from fertilisers and sewage cause eutrophication which eventually leads to algae being the dominant life form in dead zones, using up oxygen and choking the waters of other life. Dead zones continue to grow in number, up from 49 in the 1960s to 405, by 2008. The Black Sea has partially recovered from Soviet era eutrophication due to reductions in fertiliser use since the collapse of the USSR. This shows that marine ecosystems can recover if humans halt pollution.

REFERENCE: Biello, D. (2008), 'Oceanic Dead Zones Continue to Spread', *Scientific American*. http://www.scientificamerican.com/article.cfm?id=oceanic-dead-zones-spread

P. 118–119 | OCEANS FACT 26

85%: The percentage of sewage which enters the oceans untreated in East and South Asia, West and Central Africa and the South East Pacific.

COMMENT: In Europe and North America over 90% of sewage is treated before it is discharged into rivers and seas, falling to around 50% in the Mediterranean and 20% in the Caribbean. Across large parts of the developing world, however, sewage treatment is rarer. Urban populations in Asia and Africa are growing rapidly creating more megacities with highly concentrated populations but few sewage treatment facilities. Raw sewage pollutes water and leads to the spread of 'dead zones' offshore from these burgeoning cities as well as causing a threat to human health for the populations living nearby.

REFERENCE: 'The State of the Marine Environment: Trends and processes' United Nations Environment Programme (UNEP), Global Programme of Action for the Protection of the Marine Environment from Land-based Activities (GPA), 2006. http://www.gpa.unep.org/documents/global_soe_webversion_english.pdf

P. 120–121 | OCEANS FACT 27

30%: The increase in mercury levels in the Pacific Ocean between the mid 1990s and 2006.

50%: The further increase in mercury levels in the Pacific expected by 2050.

COMMENT: Mercury is deposited in the oceans by rivers and atmospheric deposition and is highly toxic to humans. Mercury levels in ocean water pose little risk, but mercury is biomagnified by the marine food web. Levels in large predatory fish such as tuna can be thousands of times higher than the levels found in ocean water. In the United States, consumption of ocean fish and shellfish accounts for 90% of human exposure to methyl mercury. Consumption of large quantities of some fish can raise mercury levels to unsafe levels. The Food Standards Agency in the UK advises that pregnant women should limit tuna consumption and that those under 16 should not eat shark, swordfish and marlin.

REFERENCE: Sunderland, E.M., Krabbenhoft, D.P., Moreau, J.W., *et al.* (2009) 'Mercury sources, distribution and bioavailability in the North Pacific Ocean—Insights from data and models' *Global Biogeochemical Cycles*, 23. || Sunderland, E. M. (2007) 'Mercury exposure from domestic and imported estuarine and marine fish in the United States seafood market', *Environmental Health Perspectives*, 115.2, pp 235–242, http://www.ncbi.nlm.nih.gov/pmc/articles/PMC1817718 || Summarised online at: http://toxics.usgs.gov/highlights/pacific_mercury.html and http://www.food.gov.uk/multimedia/faq/mercuryfish/

64,000 : The number of ships of over 400 tonnes which traversed the world's oceans in 2007, emitting 1,019 million tonnes of carbon dioxide.

COMMENT: The oceans play an ever more important role in international trade and they have become the highways of globalization. In order to feed the demand for consumer goods and exotic foods, container ships and oil tankers traverse the oceans moving goods from producers to consumers. Ocean-going ships pose an environmental threat due to the huge quantities of diesel fuel they consume. This accounts for around 3% of all global greenhouse gases.

REFERENCE: International Maritime Organisation (2009) 'World Maritime Day 2009 Climate Change: A challenge for IMO too!' http://www.imo.org/includes/blastDataOnly.asp/data_id%3D26485/WMD2009brochure.pdf

Chapter Three: The Atmosphere
P. 126–127 | ATMOSPHERE FACT 1

387 parts per million: The concentration in 2009, an increase of 23% in 50 years. Carbon dioxide levels are considered to be higher today than at any time in the last 650,000 years.

315 parts per million: The concentration of carbon dioxide in the atmosphere in 1958. This was the year continuous measurement of carbon dioxide levels began at Mauna Loa on Hawaii.

COMMENT: The rate of increase in carbon dioxide concentrations has accelerated. In the 1950s and 1960s annual increases averaged around 1 part per million per year. By the first decade of the 21st century 2.0–2.5 parts per million per year was the norm. What is most alarming is that there is as yet no sign that the rate of increase in the level of carbon dioxide in the atmosphere is even stabilizing. Carbon dioxide and other greenhouse gases emitted by human activities have enhanced the natural greenhouse effect, trapping more outgoing longwave radiation and causing global warming.

REFERENCE: NOAA, Earth System Research Laboratory Global Monitoring Division (2010), 'Trends in Atmospheric Carbon Dioxide - Mauna Loa', Dr. Pieter Tans, www.esrl.noaa.gov/gmd/ccgg/trends/ || Siegenthaler, U., Stocker, T., Monnin, E., et al. (2005) 'Stable Carbon Cycle—Climate Relationship During the Late Pleistocene' Science 310. 5752, pp. 1313–1317.

P. 128–129 | ATMOSPHERE FACT 2

450 parts per million of carbon dioxide has been identified by some scientists as a 'tipping point' for climate change, beyond which change would be dangerous and irreversible.

2042: The approximate year, at current rates, that the world will reach a concentration of 450 parts per million of carbon dioxide.

COMMENT: There is considerable debate about the tipping point at which climate change becomes irreversible and temperature rises potentially endanger global food and water supply. Levels of 560ppm are widely viewed as dangerous, although a level of 450 ppm may guarantee a global average temperature rise of +2°C which many scientists consider too much. According to climatologist James Hansen, of the NASA Goddard Institute for Space Studies, levels of 350ppm (below the 2009 level) may be enough to trigger irreversible change. He points out that in the past the poles were largely ice free at a level of 425ppm.

REFERENCE: Report of the International Scientific Steering Committee (2005) Avoiding Dangerous Climate Change. *International Symposium on the Stabilisation of greenhouse gas concentrations*, Hadley Centre, Met Office, Exeter, UK 1–3 February 2005. http://www.stabilisation2005.com/Steering_Commitee_Report.pdf

|| Lemonick, M. D. (2008), 'Beyond the Tipping Point' *Scientific American* special issue Earth 3.0, 18.4.

P. 130–131 | ATMOSPHERE FACT 3

0.02: Tonnes of carbon dioxide emitted per person by the average Somali and Afghani in 2006. Most African and many Asian nations emitted under 1 tonne of carbon dioxide per person.

56.2: Tonnes of carbon dioxide emitted per person emitted by the average Qatari, the highest level in the world in 2006. Developed and oil-rich nations emit 100–1,000 times more carbon dioxide per person than poorer developing nations.

COMMENT: High fossil fuel use increases carbon dioxide emissions. There is a strong link between a nation's wealth and its level of carbon dioxide emissions. Some oil rich nations subsidize the cost of petrol and diesel and might be accused of exacerbating the problem by encouraging heavier use of fossil fuels. A few nations have managed to break this link, for example France and Sweden emitted 5.6 and 6.2 tonnes of carbon dioxide per person respectively in 2006.

REFERENCE: United Nations Statistics Division, Millennium Indicators Database, Carbon dioxide emissions (CO2), thousand metric tons of CO2 (UNFCCC) data set, http://www.data.un.org/Default.aspx

P. 132–133 | ATMOSPHERE FACT 4

148%: The increase in concentration of methane levels in the atmosphere since 1750. The pre-industrial level of 715 parts per billion had risen to 1,774 ppb by 2005.

21: Methane has 21 times the global warming potential of carbon dioxide, making it a much more powerful greenhouse gas.

COMMENT: Global Warming Potential (GWP) measures the amount of heat trapped by the same mass of different greenhouse gases, over a set time period. Carbon dioxide has a GWP of 1 over a 100 year time span whereas methane has a GWP of 21. Molecule for molecule the methane emitted by livestock, leaking natural gas pipelines and decomposing vegetation, represents a major contributor to the global warming problem.

REFERENCE: Forster, P., Ramaswamy, V., Artaxo, P., et al. (2007) 'Changes in Atmospheric Constituents and in Radiative Forcing' in *Climate Change 2007: The Physical Science Basis, Contribution of Working Group I to the Fourth Assessment Report of the Intergovernmental Panel on Climate Change.* Cambridge and New York: Cambridge University Press.

P. 134–135 | ATMOSPHERE FACT 5

20%: Of the world's electricity consumption is simply used to keep the lights on.

75% : The average reduction in electricity consumption for lighting achieved by switching incandescent bulbs to more efficient CFL or LED lighting.

COMMENT: Old-style incandescent bulbs produce as much heat as they do light which, in most cases, is wasted energy. Switching to more efficient (and cooler) lighting saves energy, money and reduces carbon emissions. In the developing world electric lighting represents progress as under-developed regions are connected to expanding power grids. As increasing numbers of people are 'switched on' the challenge is to encourage adoption of efficient technologies.

REFERENCE: *National Geographic, Energy for Tomorrow* special issue, June 2009, ISSN 1536-6596, C. Johns (ed.).

P. 136–137 | ATMOSPHERE FACT 6

40 Billion tonnes of carbon dioxide are expected to be released due to deforestation between 2008 and 2012.

120–400 Tonnes is the amount of carbon stored by a single hectare of tropical forest.

COMMENT: While burning fossil fuels is the primary

cause of the rise in carbon dioxide levels, deforestation accounts for 18–20% of it. Cutting down living forests, which are frequently burnt, not only releases carbon dioxide directly into the atmosphere but also removes the very trees which absorb and store it. Large scale deforestation is therefore extremely foolish in a world which needs to stabilize, then reduce, its greenhouse gas emissions.

REFERENCE: Mitchell, A. W., Secoy, K., Mardas, N., *et al.* (2008). 'Forests NOW in the Fight Against Climate Change' Forest Foresight Report 1.v3 Global Canopy Programme, Oxford. http://www.globalcanopy.org/themedia/file/PDFs/Forests%20Now%20Report_Nov%2008.pdf

P. 138–139 | ATMOSPHERE FACT 7
158 Million people in the United States live in areas where local air quality failed one or more national air quality standards. In 2007 this was 52% of the US population.

COMMENT: Despite significant improvements in air quality in the United States, local air pollution is a significant problem. Cities such as Los Angeles suffer from high levels of ground level ozone, particulate matter, nitrous oxide and sulphur dioxide. High levels of this type of pollution have a significant negative effect on health, especially for the very young, the elderly and those who have pre-existing medical conditions such as asthma and heart disease. Cities often have high levels of nitrogen oxides, carbon monoxide and volatile organic compound pollution emitted by industry and transport. Sunlight promotes chemical reactions between these pollutants, breaking down the pollutants to form ozone.

REFERENCE: 'National Air Quality Status and trends through 2007', U.S. Environmental Protection Agency Office of Air Quality Planning and Standards Air Quality Assessment Division Research Triangle Park, North Carolina, EPA-454/R-08-006 November 2008. http://www.epa.gov/airtrends/2008

P. 140–141 | ATMOSPHERE FACT 8
8.1 Tonnes of carbon dioxide equivalent is the carbon footprint of the annual average US household from food consumption alone.

Livestock farming accounts for 18% of man-made greenhouse emissions: 9 per cent of all carbon dioxide, 35–40% of methane and 65% of nitrous oxide.

COMMENT: Food consumption, particularly in the developed world, is a major source of greenhouse gas emissions. Modern farming relies heavily on fossil fuels to power machinery, which emits carbon dioxide and nitrous oxide. The use of fertilisers also emits nitrous oxide which is a much more powerful greenhouse gas than carbon dioxide. Methane is released in the production of meat and dairy products. All these emissions occur on the farm, before the emissions associated with processing, packaging, transport and storage have been calculated. Each litre of milk you buy is responsible for around 1 kg of carbon dioxide emissions. A diet with a high proportion of meat is responsible for considerably larger greenhouse emissions than a vegetarian diet.

REFERENCE: Trivedi, B (2008) 'Dinner's dirty secret', *New Scientist*, 13th September 2008, 199:2673

P. 142–143 | ATMOSPHERE FACT 9
2050: The earliest that the global ozone layer could recover from depletion caused by CFCs and other pollutants.

1987: The year the Montreal Protocol on Substances that Deplete the Ozone Layer was signed.

COMMENT: Atmospheric ozone is a 'Jekyll and Hyde' gas. At ground level it is a pollutant, but high in the stratosphere it protects living things by filtering out harmful ultraviolet light. In the early 1970s and

1980s research, particularly in Antarctica, revealed the existence of a 'hole' in the stratospheric ozone layer. This hole was actually a thinning of the ozone layer caused by chlorofluorocarbons (CFCs) and other pollutants. CFCs are man-made chemical compounds (freon is an example) which were widely used as refrigerants, solvents and as aerosol propellants. The world reacted unusually quickly to this threat to the ozone layer and in 1987 signed the Montreal Protocol to progressively replace, restrict and reduce CFC emissions. In recent years, evidence from Antarctica has suggested that ozone levels are stabilising, and possibly beginning to rise, in response to reduced pollution. CFCs are such stable compounds that they persist in the environment for decades. It will be at least another generation before we will know if the ozone layer has recovered.

REFERENCE: British Antarctic Survey, Natural Environment Research Council, 'Science Briefing, 2008, Ozone', http://www.antarctica.ac.uk/press/journalists/resources/science/the_ozone_hole_2008.pdf

P. 144–145 | ATMOSPHERE FACT 10
20%: Decline in sunlight since the 1970s at Guangzhou in China, as a result of rising levels of industrial pollution.

3–4%: Decline in sunlight over much of India and China, per decade, since the 1950s.

COMMENT: Declining levels of sunlight at ground level are often referred to as 'global dimming'. The dimming of the sun results from Atmospheric Brown Clouds – most famously the 'Asian Brown Cloud' observed over south and east Asia. Brown clouds consist of black carbon (soot), ash, nitrogen dioxide, sulphur dioxide and dust released into the lower atmosphere by industry, farming and transport. Brown cloud pollution not only severely affects human health, for instance increasing the prevalence of respiratory illness, but also is linked to changing patterns of rainfall, particularly the south Asian monsoon, falling crop yield and increased melting of Himalayan glaciers.

REFERENCE: Ramanathan, V., Agrawal, M., Akimoto, H., *et al.* (2008), 'Atmospheric Brown Clouds: Regional Assessment Report with Focus on Asia' United Nations Environment Programme, Nairobi, Kenya. http://www.unep.org/pdf/ABCSummaryFinal.pdf

P. 146–147 | ATMOSPHERE FACT 11
One 500 megawatt coal-fired power station produces the same annual carbon dioxide emission as around 600,000 cars.

COMMENT: Coal is a very abundant fossil fuel, but it is also a much dirtier fuel than either oil or gas. In 2004–05 the United States planned to build around 150 new coal-fired power stations. As a result of environmental campaigns, legal challenges and changes to emissions regulations, most of these will not be built. Elsewhere in the world, however, coal is likely to be the backbone of an expansion in electricity generation for some time to come. Both India and China have plans to build hundreds of new coal fired plants over the next few decades. Some of these plants will be more efficient and cleaner than those of the past, but they will still release greenhouse gases and other harmful products.

REFERENCE: Union of Concerned Scientists, 'The Costs of Coal Across the fuel cycle: from mine to smokestack', http://www.ucsusa.org/clean_energy/technology_and_impacts/impacts/the-costs-of-coal.html || Jowitt, J. (2008), *The Guardian*, 3 September 2008, 'Coal Plans go Up in Smoke', http://www.guardian.co.uk/environment/2008/sep/03/activists.fossilfuels

P. 148–149 | ATMOSPHERE FACT 12
Sulphur Dioxide, Carbon Dioxide, Nitrogen Oxide,

Carbon Monoxide, particulate matter, unburned hydrocarbons, ash, arsenic, lead, cadmium and mercury are all released into the atmosphere when coal is burnt.

COMMENT: Coal's problem is not just the carbon dioxide it emits. A 500 megawatt coal plant burning 1.3 million tonnes of coal per year emits around 10,000 tonnes of both sulphur dioxide and nitrogen oxide and up to 500 tonnes of particulate matter. Many coal fired power stations achieve an efficiency of only 25–35%, that is only about 1/3 of the energy available in each tonne of coal is actually converted into electrical power – the rest is lost as waste heat. Increasingly, scientists and environmentalists argue that while coal is cheap and abundant it is also dirty and inefficient.

REFERENCE: Union of Concerned Scientists 'How Coal Works', website http://www.ucsusa.org/clean_energy/ technology_and_impacts/impacts/the-costs-of-coal.html

P. 150–151 | ATMOSPHERE FACT 13
3830: The global warming potential of HFC-134a, a gas used as an alternative to ozone depleting CFCs.
340%: The increase in concentration of HFC-134a in the atmosphere between 1998 and 2005.

COMMENT: Following the 1987 Montreal Protocol many ozone depleting CFCs (chlorofluorocarbons) were phased out to be replaced by less harmful alternatives. An example is the replacement of CFC-12 with HFC-134a (hydrofluorocarbon) in air conditioners. HFC-134a does not damage the ozone layer, but it is a powerful greenhouse gas. Using the standard measure of Global Warming Potential (GWP) each kilogram of HFC-134a has 3,830 times the warming potential of one kilogram of carbon dioxide over a 20 year period. The concentration of HFC-134a in the atmosphere increased from 8 parts per trillion in 1998 to 35 parts per trillion in 2005. Rapid rises in this and similar powerful greenhouse gases are increasingly a cause for concern.

REFERENCE: Forster, P., Ramaswamy, V., Artaxo, P., et al. (2007) 'Changes in Atmospheric Constituents and in Radiative Forcing', in S. Solomon, D. Qin, M. Manning, et al. (eds.) *Climate Change 2007: The Physical Science Basis, Contribution of Working Group I to the Fourth Assessment Report of the Intergovernmental Panel on Climate Change*, Cambridge and New York: Cambridge University Press.

P. 152–153 | ATMOSPHERE FACT 14
75%: The approximate increase in the percentage of category 4 and 5 tropical cyclones since 1970.

COMMENT: Tropical cyclones (also called hurricanes and typhoons) might be expected to become more powerful in a warming world. Tropical cyclones are highly energetic storms which derive their energy from warm ocean waters. If sea surface temperatures rise due to global warming, then tropical cyclones should have more energy and be larger and more destructive. Tropical cyclones only form over oceans where the temperature exceeds 26°C. Warmer oceans, therefore, may mean an expansion of the area vulnerable to tropical cyclones. There is an ongoing debate over the reliability of the evidence of increased tropical cyclone intensity. It is an important debate because of the destructive power of major cyclones such as Hurricane Katrina (2005) and Cyclone Nargis (2008).

REFERENCE: Trenberth, K., Jones, P., Ambenje, P., et al. (2007), 'Observations: Surface and Atmospheric Climate Change' in *Climate Change 2007: The Physical Science Basis, Contribution of Working Group I to the Fourth Assessment Report of the Intergovernmental Panel on Climate Change*. Cambridge and New York: Cambridge University Press.

P. 154–155 | ATMOSPHERE FACT 15
3–4°C: The warming experienced in Western Canadian and Alaskan Arctic over the last 50 years. The Arctic has warmed twice as fast as the global average in the past few decades and this trend is expected to continue in the 21st century.

COMMENT: The fragile Arctic tundra environment appears to be changing more rapidly as a result of global warming than any other global environment. In a major Scientific study in 2004, the Arctic Climate Impact Assessment warned that over the course of the 21st century Arctic oceans could warm by 7°C and land areas by 3–5°C. Warming on this scale would transform the landscape and ecosystems of the Arctic, spelling disaster for sea-ice dependent animals such as ringed seals and polar bears. The Arctic ecosystem relies on intensely cold winters, to which it is well adapted. The ACIA warms that winters could warm by 4–7°C by 2100. Arctic summer sea-ice declined in extent by 15–20% between 1970 and 2005.

REFERENCE: Arctic Council, 'Arctic Climate Impact Assessment' (2004) presented at Fourth Arctic Council Ministerial Meeting, Reykjavik. Cambridge and New York: Cambridge University Press.http://www.acia.uaf.edu/

P. 156–157 | ATMOSPHERE FACT 16
23%: The drop in number of species in grasslands in the UK caused by nitrogen deposition.
17: The average deposition rate of nitrogen in kg N Ha-1 yr -1 (kilograms of nitrogen per hectare per year) over the UK from both agricultural and energy production sources.

COMMENT: There are two types of acid deposition – dry deposition when the NO and NO_2 falls back down to earth as a gas and wet deposition when the gases combine with precipitation to form nitric acid. Both have a severe impact on ecosystems. Plant species are reduced by two processes, the top down process and the bottom up process, caused by these two different types of deposition. In the top down process the dry deposition interferes with the ability of the plant to photosynthesize and so eventually the plant dies. In the bottom up process, caused by wet deposition, the nutrients are washed from the soil and the plant no longer has access to the vital nitrogen, phosphorus, and potassium and so also dies.

REFERENCE: Stevens, C., Dise, N., Mountford, J., et al (2004) ' Impact of Nitrogen Deposition on the Species Richness of Grasslands'. *Science*, 303, 1876–1879.

P. 158–159 | ATMOSPHERE FACT 17
40%: The percentage of oak stands showing severe defoliation in Germany.
25%: The percentage of all trees in 27 European countries that also showed severe defoliation.

COMMENT: Although not everyone agrees, some scientists have demonstrated a link between changes in the soil chemistry of Central European forests and forest defoliation. Due to wet deposition, the soil not only became more acidic, but also showed an increase in the amount of nitrogen. This encouraged the free root biomass of the trees to shift to the upper soil layers and for the total root biomass to become smaller. There also appeared to be significant growth of the forests above ground which meant that the ratio of root mass to above ground growth decreased, leaving the trees more susceptible to drought and thus further stress leading to the observed defoliation. This defoliation was then made worse by the additional direct effects of dry deposition on the leaves of forests in Central Europe.

REFERENCE: BML, (1994). *Waldzustandsbericht der Bundesregierung 1994 – Ergebnisse der Waldschadenserhebung.* || Kladtke, J. (1995), *UBA-Texte 28,* 120–130. Cited in Matzner, E. and Murach, D. (1995) 'Soil changes induced by air pollutant deposition and their implication for forests in Central Europe' *Water, Air and Soil Pollution* 85: 63–76 || Matzner, E. and Murach, D. (1995) ' Soil changes induced by air pollutant deposition and their implication for forests in Central Europe'

Water, Air and Soil Pollution 85: 63–76 || Ulrich, B. (1984) 'Effects of air pollution of forest ecosystems and waters: the principles demonstrated at a case study in Central Europe' *Atmospheric Environment* 18:3 pp 621–628. || Wesselink, L., Meiwes, K., Matzner, E., *et al*. (1995) 'Long-term changes in water and soil chemistry in spruce and beech forests, Solling, Germany', *Environmental Science & Technology* 29, 51–58.

P. 160–161 | ATMOSPHERE FACT 18

316,000: The number of lost years due to people dying prematurely due to air pollution in cities in France in one year .
55%: The number of these premature deaths attributable to air pollution caused by transport.

COMMENT: Concern over the quality of air in urban areas has its roots in the Great Smogs in London during the 1950s. Indeed it was the Great Smog of 1952 which is estimated to have killed between 4,000 and 12,000 people and which led to the Clean Air Acts in the UK. Although the problem of burning coal in urban areas was solved, the huge rise in car ownership gave rise to new threats to human health. The use of lead in petrol as an anti-knocking agent in cars and recent increases in the levels of carbon monoxide, nitrous oxide, volatile organic compounds and particulate matter from car exhausts all have deleterious effects on human health. Studies of hospital admissions for respiratory and cardiovascular cases, found that out of all the incidents of chronic bronchitis in adults and bronchitis episodes in children, there were 40,000 deaths per year attributable to air pollution. As well as these deaths, there were more than 25,000 new cases of chronic bronchitis and more than 290,000 episodes of bronchitis and more than 0·5 million asthma attacks. It appears therefore that we might have reduced one form of urban air pollution only to replace it with another.

REFERENCE: CITEPA, 2001 Inventaire des emissions de l'air en France—format Secten (air emission inventory in France). Convention report 9/98 || Künzli, N., Kaiser, R., Medina, S., *et al*. (2000) 'Public-health impact of outdoor and traffic-related air pollution: a European assessment'. *Lancet*, 356, 795–801 || Nicolas, J.P., Duprez, F., Durand, S., *et al*. (2005) ' Local impact of air pollution: lessons from recent practices in economics and in public policies in the transport sector'. *Atmospheric Environment*, 39, 2475–2482.

P. 162–163 | ATMOSPHERE FACT 19

300%: The increase in the number of child deaths from asthma in the US in the period 1979 to 1996.
80%: The percentage of asthma cases in the US believed to be caused by a reaction to allergens and indoor pollution.

COMMENT: According to the Institute of Medicine (2000), asthma is the most common chronic health problem in children and one of the most common health complaints in the US. Unlike many other chronic diseases, asthma rates appear to be increasing particularly among disadvantaged inner-city minority children. Disease rates are highest among African-Americans, Hispanics, and populations in urban inner cities. The link between increased urban air pollution and increased recorded cases of asthma has been well established in numerous studies.

REFERENCE: Breysse, P., Buckley, T., Williams, D., *et al*. (2005) 'Indoor exposures to air pollutants and allergens in the homes of asthmatic children in inner-city Baltimore' *Environmental Research*, 98, 167–176. || Institute of Medicine (2000). *Clearing the Air: Asthma and Indoor Air Exposures*. Washington, DC: National Academy Press.

P. 164–165 | ATMOSPHERE FACT 20

4,000: The number of toxic chemicals in environmental tobacco smoke.
30%: Increase in the risk of death from heart disease due to passive smoking.

COMMENT: Burning cigarettes emit two types of smoke: mainstream smoke, which is the smoke directly inhaled into the smoker's lungs, and sidestream smoke, which is the smoke emitted into the air from the burning cigarette between puffs. It is believed that environmental tobacco smoke is about 85% sidestream and 15% exhaled mainstream smoke. In addition to the 4,000 toxic chemicals found in environmental tobacco smoke there are also at least 40 carcinogens. Many of these toxic constituents are found in higher concentrations in sidestream (i.e. passive smoke) than in mainstream smoke. For example, sidestream smoke contains about five times as much carbon monoxide (which decreases the ability of hemoglobin to carry oxygen to the tissues), three times as much benzopyrene (a tumor and plaque producing compound), and fifty times as much ammonia (an eye and respiratory irritant) as smoke inhaled directly from a cigarette.

As tobacco smoke can persist in indoor environments for many hours after the cessation of smoking, there is considered to be a causal relationship between parental smoking and acute lower respiratory illness in young children.

REFERENCE: Otsuka, R., Watanabe, H., Hirata, K., *et al*. (2001) 'Acute effects of Passive smoking on the coronary circulation in healthy young adults' *Jama*, 286, 1–11. || Strachan, D. and Cook, D. (1997) 'Health effects of passive smoking Parental smoking and lower respiratory illness in infancy and early childhood' *Thorax* 1997, 52: 905–914. || Taylor E., Johnson, D. and Kazemi, H. (1992) 'Environmental Tobacco Smoke and Cardiovascular Disease. A Position Paper From the Council on Cardiopulmonary and Critical Care, American Heart Association' *Circulation*, 6, 699–702.

P. 166–167 | ATMOSPHERE FACT 21

60: The number of cigarettes that you would need to smoke a day to equal the average air pollution in Mexico City.
100%: The number of days that the PM$_{10}$ value exceeds the WHO guideline. This refers to particulate matter that can be inhaled.

COMMENT: Mexico City has one of the biggest air pollution problems in the world. There are 2.5 million registered vehicles, but the figure is probably higher if unregistered vehicles are included. Buses, cars, taxis and motorbikes create the majority of the pollution. The rest of the pollution comes from the 30,000 factories situated in Mexico City. These produce mainly heavy metal contamination as metal is smelted, but vast quantities of PM$_{10}$ are also emitted from factories such as the PEMEX oil refinery. The pollution levels are made worse by temperature inversions which trap the pollution at ground level. A temperature inversion occurs when air is trapped at ground level because the overlying air is warmer than the air at ground level These inversions are caused in part because Mexico City is located at a high altitude and surrounded by volcanic mountains. Due to its high altitude, the oxygen level is lower so the concentration of pollutants such as carbon monoxide are higher than they would be at a lower altitude. In the afternoon when the temperature inversion is at its highest, a photochemical smog envelops the city.

REFERENCE: Ward, P. (1998). *Mexico City*, 2nd Edition. London : Belhaven

P. 168–169 | ATMOSPHERE FACT 22

1.4 million: The number of premature deaths every year in Africa caused by the use of dirty biomass fuels to meet

families' daily energy needs.

11.2%: The percentage of lower respiratory infections (ALRI) of the total burden of disease in Africa, second only to the continent's HIV/AIDS pandemic.

COMMENT: In addition to the problems of urban air pollution, there is now increasing concern over the effects of indoor air pollution, particularly in poorer countries. The inefficient burning of kerosene and charcoal leads to very high concentrations of indoor air pollutants such as carbon monoxide (CO) and fine particulate matter ($PM_{2.5}$). Of particular concern is the effect on women and children who may spend large amounts of time in the kitchen. Indoor air pollution from biomass fuels has been linked to increased rates of pneumonia in children and chronic obstructive pulmonary disease among adult women. Studies have also suggested that there is evidence of links to eye disease, tuberculosis and low birth weight. According to the World Health Organization, low birth weight and tuberculosis are among the top ten contributors to the total disease burden in Africa.

REFERENCE: Lin, H., Ezzati, M. and Murray M. (2007) 'Tobacco smoke, indoor air pollution and tuberculosis: a systematic review and meta-analysis' *PLoS Medicine* 4(1), 173–189 || Pennise, D., Brant, S., Agbeve, S., *et al* (2009). 'Indoor air quality impacts of an improved wood stove in Ghana and an ethanol stove in Ethiopia' *Energy for Sustainable Development*, 13, 71–76 || World Health Organization (2008) 'The global burden of disease: 2004 update' Geneva : World Health Organization. http://www.who.int/healthinfo/global_burden_disease/GBD_report_2004update_full.pdf

P. 170–171 | ATMOSPHERE FACT 23

$80: The average cost of a clean cook stove sold in Ethiopia that reduces the CO produced by such indoor stoves by 76% and so below the WHO guidelines for good health.

$140: The average GDP per capita of Ethiopia in 2009. This effectively means that it is unaffordable by the vast majority of families in Ethiopia.

COMMENT: One of the key problems in providing secure energy as well as clean energy in poorer countries is the cost of such 'westernised' approaches. Clean cook stoves would undoubtedly save lives, but unless there are fresh injections of capital into the aid budgets from richer countries, the prospect of a majority of households in countries such as Ethiopia having cleaner and safer cooking technology seems a long way off.

REFERENCE: Pennise, D., Brant, S., Agbeve, S., *et al*. (2009).' Indoor air quality impacts of an improved wood stove in Ghana and an ethanol stove in Ethiopia' *Energy for Sustainable Development*, 13, 71–76.

P. 172–173 | ATMOSPHERE FACT 24

18,448,200: The number of registered residential vehicles in California.

In California, the damage from urban air pollution to the capacity of children's lungs to obtain oxygen is 4.5 times greater than the damage from passive smoking.

COMMENT: California not only has one of the lowest levels of smoking in the US at 19.5%, it also had one of the biggest relative declines in smoking (22.9%) between 1985 and 1993. It was therefore hoped that the effects of this reduction in smoking would lead to a corresponding decline in the damage to developing lung function in children. Unfortunately, it seems that one pollutant has merely been replaced by another. The acute health consequences of breathing polluted air range from an increase in cardio-respiratory morbidity and mortality to an increase in respiratory symptoms and decrease in lung function. While Californians might have reduced the effects of passive smoking, their reliance on the motor car has created new and deadlier effects – particularly

for children who tend to spend more time outdoors. Unlike the lower rate of smoking, car ownership in California is one of the highest in the US with over 13% of all registered vehicles in the US being registered in California. It appears that even such health conscious states such as California have a long way to go before their children can grow up free of the risk of the acute health problems caused by air pollution.

REFERENCE: Gauderman, W., Connell, R., Gilliland, F., *et al.* (2000) 'Association between Air Pollution and Lung Function Growth in Southern California Children' *American Journal of Respiratory Critical Care* 162 1383–1390 || Purvis, C. (1997) 'Auto Ownership in the San Francisco Bay Area: 1930 – 2010' MTC http://www.mtc.ca.gov/maps_and_data/datamart/forecast/ao/aopaper.htm || Shopland, D., Hartman, A., Gibson, J., *et al.* (1996) 'Cigarette Smoking Among U.S. Adults by State and Region: Estimates from the Current Population Survey' Journal of the *National Cancer Institute*, 88, 1748–1758

P. 174–175 | ATMOSPHERE FACT 25

105 ppb: The maximum suggested short-term concentration of NO_2 that can be tolerated before there are severe health consequences.

Research carried out in homes in central London predicted that in a one hour period the predicted NO_2 levels were 600 ppb.

COMMENT: As well as health issues in poorer countries, richer countries such as the UK also have health issues associated with indoor air pollution. Gas cooking is considered to be the dominant source of NO_2 in UK homes. NO_2 concentrations in the kitchen during cooking consistently exceeded the suggested short-term UK guideline of 105 ppb. This suggests that even with a reduction in urban air pollution through sustainable techniques, people are still at risk from air pollution.

REFERENCE: Dimitroulopoulou, C., Ashmore, M., Hill, M., *et al.* (2006) 'INDAIR: A probabilistic model of indoor air pollution in UK homes' *Atmospheric Environment*, 40, 6362–6379 || Ross, D., (1996) 'Continuous monitoring of NO_2 and CO, temperature and humidity in UK' in *Proceedings of Indoor Air* 1996, Japan, 1, 513–518.